Selected Titles in This Series

D1684601

667 **Ethan Akin,** Simplicial dynamical systems, 1999

666 **Mark Hovey and Neil P. Strickland,** Morava K-theories and localisation, 1999

665 **George Lawrence Ashline,** The defect relation of meromorphic maps on parabolic manifolds, 1999

664 **Xia Chen,** Limit theorems for functionals of ergodic Markov chains with general state space, 1999

663 **Ola Bratteli and Palle E. T. Jorgensen,** Iterated function systems and permutation representation of the Cuntz algebra, 1999

662 **B. H. Bowditch,** Treelike structures arising from continua and convergence groups, 1999

661 **J. P. C. Greenlees,** Rational S^1-equivariant stable homotopy theory, 1999

660 **Dale E. Alspach,** Tensor products and independent sums of $\mathcal{L}_p$-spaces, $1 < p < \infty$, 1999

659 **R. D. Nussbaum and S. M. Verduyn Lunel,** Generalizations of the Perron-Frobenius theorem for nonlinear maps, 1999

658 **Hasna Riahi,** Study of the critical points at infinity arising from the failure of the Palais-Smale condition for n-body type problems, 1999

657 **Richard F. Bass and Krzysztof Burdzy,** Cutting Brownian paths, 1999

656 **W. G. Bade, H. G. Dales, and Z. A. Lykova,** Algebraic and strong splittings of extensions of Banach algebras, 1999

655 **Yuval Z. Flicker,** Matching of orbital integrals on $GL(4)$ and $GSp(2)$, 1999

654 **Wancheng Sheng and Tong Zhang,** The Riemann problem for the transportation equations in gas dynamics, 1999

653 **L. C. Evans and W. Gangbo,** Differential equations methods for the Monge-Kantorovich mass transfer problem, 1999

652 **Arne Meurman and Mirko Primc,** Annihilating fields of standard modules of $\mathfrak{sl}(2, \mathbb{C})^{\sim}$ and combinatorial identities, 1999

651 **Lindsay N. Childs, Cornelius Greither, David J. Moss, Jim Sauerberg, and Karl Zimmermann,** Hopf algebras, polynomial formal groups, and Raynaud orders, 1998

650 **Ian M. Musson and Michel Van den Bergh,** Invariants under Tori of rings of differential operators and related topics, 1998

649 **Bernd Stellmacher and Franz Georg Timmesfeld,** Rank 3 amalgams, 1998

648 **Raúl E. Curto and Lawrence A. Fialkow,** Flat extensions of positive moment matrices: Recursively generated relations, 1998

647 **Wenxian Shen and Yingfei Yi,** Almost automorphic and almost periodic dynamics in skew-product semiflows, 1998

646 **Russell Johnson and Mahesh Nerurkar,** Controllability, stabilization, and the regulator problem for random differential systems, 1998

645 **Peter W. Bates, Kening Lu, and Chongchun Zeng,** Existence and persistence of invariant manifolds for semiflows in Banach space, 1998

644 **Michael David Weiner,** Bosonic construction of vertex operator para-algebras from symplectic affine Kac-Moody algebras, 1998

643 **Józef Dodziuk and Jay Jorgenson,** Spectral asymptotics on degenerating hyperbolic 3-manifolds, 1998

642 **Chu Wenchang,** Basic almost-poised hypergeometric series, 1998

641 **W. Bulla, F. Gesztesy, H. Holden, and G. Teschl,** Algebro-geometric quasi-periodic finite-gap solutions of the Toda and Kac-van Moerbeke hierarchies, 1998

640 **Xingde Dai and David R. Larson,** Wandering vectors for unitary systems and orthogonal wavelets, 1998

639 **Joan C. Artés, Robert E. Kooij, and Jaume Llibre,** Structurally stable quadratic vector fields, 1998

(*Continued in the back of this publication*)

Simplicial Dynamical Systems

MEMOIRS
of the
American Mathematical Society

Number 667

Simplicial Dynamical Systems

Ethan Akin

July 1999 • Volume 140 • Number 667 (first of 4 numbers) • ISSN 0065-9266

American Mathematical Society
Providence, Rhode Island

1991 *Mathematics Subject Classification.*
Primary 54H20, 58F10, 34C35.

Library of Congress Cataloging-in-Publication Data

Akin, Ethan, 1946–
Simplicial dynamical systems / Ethan Akin.
p. cm. — (Memoirs of the American Mathematical Society, ISSN 0065-9266 ; no. 667)
"July 1999, volume 140, number 667 (first of 4 numbers)."
Includes bibliographical references and index.
ISBN 0-8218-1383-8
1. Topological dynamics. 2. Stability. 3. Differentiable dynamical systems. I. Title. II. Series.
QA3.A57 no. 667
[QA611.5]
510 s—dc21
[514]

99-14982
CIP

Memoirs of the American Mathematical Society

This journal is devoted entirely to research in pure and applied mathematics.

Subscription information. The 1999 subscription begins with volume 137 and consists of six mailings, each containing one or more numbers. Subscription prices for 1999 are $448 list, $358 institutional member. A late charge of 10% of the subscription price will be imposed on orders received from nonmembers after January 1 of the subscription year. Subscribers outside the United States and India must pay a postage surcharge of $30; subscribers in India must pay a postage surcharge of $43. Expedited delivery to destinations in North America $35; elsewhere $130. Each number may be ordered separately; *please specify number* when ordering an individual number. For prices and titles of recently released numbers, see the New Publications sections of the *Notices of the American Mathematical Society.*

Back number information. For back issues see the *AMS Catalog of Publications.*

Subscriptions and orders should be addressed to the American Mathematical Society, P. O. Box 5904, Boston, MA 02206-5904. *All orders must be accompanied by payment.* Other correspondence should be addressed to Box 6248, Providence, RI 02940-6248.

Memoirs of the American Mathematical Society is published bimonthly (each volume consisting usually of more than one number) by the American Mathematical Society at 201 Charles Street, Providence, RI 02904-2294. Periodicals postage paid at Providence, RI. Postmaster: Send address changes to Memoirs, American Mathematical Society, P. O. Box 6248, Providence, RI 02940-6248.

10 9 8 7 6 5 4 3 2 1 04 03 02 01 00 99

For Jean
Just the facts, Ma'am.

Table of Contents

Abstract

A simplicial dynamical system is a simplicial map $g : K^* \to K$ where K is a finite simplicial complex triangulating a compact polyhedron X and K^* is a proper subdivision of K, e.g. the barycentric or any further subdivision. The dynamics of the associated piecewise linear map $g : X \to X$ can be analyzed by using certain naturally related subshifts of finite type. Any continuous map on X can be C^0 approximated by such systems. Other examples yield interesting subshift constructions.

Key Words and Phrases: simplicial dynamics, subshifts of finite type, basic sets, sample path spaces, Markov measures, hyperbolicity.

A.M.S. Subject Classifications (1991): 54H20, 58F10, 34C35

Introduction

Let $f : X \to X$ be a continuous map with X a compact metric space. We define a dynamical system by iterating such a function on X. A computational analysis of the system is an attempt to describe its behavior by using a finite set of data points with the individual points themselves subject to round off error.

This is a well studied problem, see especially Hsu (1987). The most direct approach is to choose a finite subset F of X, regarding it as a finite approximation of the space. For each point $z \in F$ choose $g(z)$, a point of F close to $f(z)$, so that the map g on F is regarded as an approximation of f on X. In effect, we apply f to each point of F and then round off to obtain values in F. Alternatively, we can choose a finite cover F of X by small, nonempty subsets of X, "cells" or "pixels". For each set $z \in F$ the image $f(z)$ meets one or more elements of F. This defines a relation on the finite set F which we regard as the approximation to f, see Hsu (1987), Akin (1993) Chapter 5 and Osipenko (to appear). In this monograph we introduce a third mode of approximating the function f using finite data.

Consider the simpler problem of sketching the graph of a real-valued function defined on a bounded interval. We partition the domain by small subintervals using a finite, increasing sequence of points. At each such point we evaluate the function and so compute, up to roundoff, the corresponding sequence of points on the graph. If we connect successive points by a line segment then we get the graph of a piecewise linear function which approximates the original. This function obtained by linear interpolation really carries no more information than did the finite sequence of points from which it was determined. However, it has the advantage that it is the same sort of object as that which we intended to study. It is a real-valued function on the original domain, an especially simple one because it is piecewise linear.

In our original problem we will restrict to the case where X is a compact polyhedron and we will approximate the function f on X by special simplicial maps.

We will use as references: Rourke and Sanderson (1972), Hudson (1969) and Stallings (1968). These describe a polyhedron as a special compact subset of a Euclidean space. Our results easily extend to objects in the PL category, that is, compact metric spaces equipped with a PL structure, a

Received by the editor October 30, 1996.

class of p.l. compatible homeomorphisms to polyhedra. From there they apply to smooth manifolds.

The simplices of a simplicial complex K triangulating a polyhedron X are subsets of X. Furthermore, each has a linear structure as a closed convex subset of the ambient Euclidean space.

Now suppose that a triangulation K of X consists of a small simplices. That is, the mesh of K is small, where the mesh is the maximum among the diameters of the simplices z of K. Let K' be the barycentric subdivision of K. Each simplex z of K has a regular neighborhood $N(z, K')$ the union of the simplices of K' which intersect z. Think of the z's or the $N(z, K')$'s as the cells of our second approximation method.

The image set $f(z)$ may stretch and bend in X, usually encountering many simplices of K. However, if K^* is a fine enough subdivision of K then for each $z^* \in K^*$ $f(z^*)$ is contained in the interior of $N(z, K')$ for some $z \in K$. Furthermore, it is possible to estimate how small the mesh of K^* has to be so that this condition holds.

If the image of each z^* is contained in the interior of some $N(z, K')$ then we denote the smallest such z by $s_f(z^*)$. It is easy to check that if z_1^* is a face of z^* in K^* then $s_f(z_1^*)$ is a face of $s_f(z^*)$ in K. By mapping the barycenter of z^* to the barycenter of $s_f(z^*)$ we define a simplicial map $g : K^{*\prime} \to K'$. If K^* was in fact a subdivision of K' then we have obtained from f a simplicial map $g : L^* \to L$ where $L^* = K^{*\prime}$ is a subdivision of $L = K'$. The vertices of L^* correspond to the finite set of points in our first approximation method. A simplicial map is completely determined by linearity once the vertex values are known. This is the analogy with our piecewise linear curve sketch.

A subdivision K^* of a simplicial complex K is called a *proper* subdivision if no simplex of K^* meets disjoint simplices of K. This is a very mild condition. If K^* is any subdivision of the barycentric subdivision of K then it is a proper subdivision of K. By a *simplicial dynamical system* on a polyhedron X we mean a triangulation K of X, a proper subdivision K^* of K and a simplicial map g from K^* to K. We use the same symbol for both the mapping of finite sets of simplices, $g : K^* \to K$, and for the underlying piecewise linear mapping g on X. We described above a way of approximating the dynamical system f by a simplicial dynamical system. The approximating map g on X will be called a *p.l. roundoff map* for f.

This monograph analyzes the behavior of simplicial dynamical systems.

Because $K^* \neq K$ we cannot directly iterate the simplicial map $g : K^* \to K$. If $z^* \in K^*$ then $z = g(z^*) \in K$ and so usually contains many simplices of K^*. Thus, we can define a multiple valued function, or relation, G^* on the finite set K^*: z_1^* is G^* related to z^* if $z_1^* \subset g(z^*)$. There is a similar relation G on K: z_1 is G related to z if $z_1 \subset g(z)$. The recurrence properties of the p.l. map g on X can be computed by using the finite relations G^* and G. We introduced these systems as approximations to general continuous maps, but the association with shift maps leads to other applications as we will later see.

As an approximation procedure p.l. roundoff, like linearization, consists of an easy part and a hard part. The easy part is setting up the approximation and analyzing the approximating object. After all if this part were not tractable the approximation would be pointless. This easy part is what the rest of this little book does in detail. Before summarizing the individual chapters, we will say a few words about the hard part: getting back to the original system from the approximate one.

The p.l. roundoff map g approximates f in the C^0 sense. To be precise, given f and $\epsilon > 0$ there exist $\delta, \delta^* > 0$ so that if $\text{mesh}(K) < \delta$ and the $\text{mesh}(K^*) < \delta^*$ where K^* is a subdivision of K', then the simplex association $s_f(z^*) \in K$ is defined for all $z^* \in K^*$ and the p.l. roundoff map g on X satisfies $d(f(x), g(x)) \leq \epsilon$ for all $x \in X$. Thus, in general, these methods detect properties which are robust enough to be preserved by C^0 approximation, see Akin (1993) Chapter 7. For this reason we focus our attention below on chain recurrence and basic sets. By imposing special conditions on f one can go farther. If g is ϵ close to f then a g orbit is an ϵ chain for f, also called an ϵ pseudo-orbit for f. By imposing hyperbolicity assumptions on f and using shadowing theorems one can sometimes show that the g orbits approximate true f orbits.

* *

Chapter 1, Chain Recurrence and Basic Sets: We review from Akin (1993) some of the notation and results for dynamics of a closed relation F on a compact metric space X. Of greatest importance is the *chain recurrent set* $|\mathcal{C}F|$ and the *basic sets* contained therein, i.e. the $\mathcal{C}F \cap (\mathcal{C}F)^{-1}$ equivalence classes, also called the chain components. In particular, we describe from Miller and Akin (1998) results comparing the recurrence properties of F with those of the two-sided shift homeomorphism s_F and the one-sided shift map s_F^+ on the associated sample path spaces $X_F \subset X^{\mathbf{Z}}$ and $X_F^+ \subset X^{\mathbf{Z}_+}$, respectively. These have two different applications.

If F is itself a continuous map on X then s_F^+ is conjugate to F itself and s_F is the shift homeomorphism on the inverse limit of the system $\dots X \xrightarrow{F} X \xrightarrow{F} X$. On the other hand, if X is a finite set, thought of as an alphabet, then F is the set of admissible dipthongs and s_F and s_F^+ are the associated subshifts of finite type.

Of special interest is the *two alphabet case.* Given two finite sets A^* and A suppose that $g : A^* \to A$ is a map and $J : A^* \to A$ is a relation. The relations $G^* = J^{-1} \circ g$ on A^* and $G = g \circ J^{-1}$ on A have closely related dynamics. Thus, when A^* is a much larger set than A we can use G to study G^*.

When f is a homeomorphism on X a *pre-decomposition* $\tilde{\mathcal{F}}$ for f is a finite collection of closed, nonempty, invariant subsets of X such that the positive and negative limits sets of any orbit sequence for f are each contained in elements of $\tilde{\mathcal{F}}$. By concatenating we obtain the associated decomposition which is a pre-decomposition by pairwise disjoint sets. We review from Akin (1993) how the basic sets are obtained from decompositions.

Chapter 2, Simplicial Maps and Their Local Inverses: We review the definitions and notations for simplicial complexes (always assumed finite) and the compact polyhedra they triangulate. When a complex K triangulates a polyhedron X, we write $X = |K|$ and define on X a metric d_K by using the l^1 norm to compare barycentric coordinates.

Between complexes K^* and K, a simplicial map $g : K^* \to K$ is a map of finite sets with special properties associated with the complex structure. e.g. g preserves incidence. There is an associated piecewise linear map, denoted $g : |K^*| \to |K|$, which is linear on each simplex of K^*.

If $g : K^* \to K$ is a simplicial map and z^* is a simplex of K^* with $z = g(z^*)$ in K then the dimensions satisfy dim $z \leq$ dim z^*. We call z^* *degenerate* if the inequality is strict and so z^* is *nondegenerate* when dim $z =$ dim z^*.

If z^* is nondegenerate then the restriction $g : z^* \to z$ is a linear isomorphism and so admits an inverse map $\overline{g}_{z^*} : z \to z^*$. When z^* is degenerate we can define a family of one-sided inverse maps. For each vertex v of z choose a point of $z^* \cap g^{-1}(v)$. The set of such choices is a convex cell we denote T^{z^*}. So z^* is nondegenerate when T^{z^*} is a single point. For each $t \in T^{z^*}$ we define a map $\overline{g}_{z^*,t} : z \to z^*$ by extending linearly from the vertex choices.

Suppose that K^* is a subdivision of K. We call K^* a *proper subdivision* of K if no simplex of K^* meets disjoint simplices of K. For example, if K' is the barycentric subdivision of K then K' and any further subdivision K^* of K' are proper for K.

A *simplicial dynamical system* is a simplicial map $g : K^* \to K$ where K^* is a proper subdivision of K. The main result of this chapter is:

Theorem: *Let $g : K^* \to K$ be a simplicial dynamical system. For $z^* \in K^*$ and $t \in T^{z^*}$ the maps $\overline{g}_{z^*,t} : g(z^*) \to z^*$ contract the metric d_K uniformly. That is, the supremum of the Lipschitz constants is less than 1.*

Chapter 3, The Shift Factor Maps for a Simplicial Dynamical System: We define the maps which are used to study a simplicial dynamical system $g : K^* \to K$. Since K and its proper subdivision K^* triangulate a common polyhedron X, the associated p.l. map g defines a topological dynamical system on X.

First, let $\hat{K}$ be the union of the pairs $\{z^*\} \times T^{z^*}$ as z^* varies over K^* with $\hat{p} : \hat{K} \to K^*$ the projection defined by $(z^*, t) \mapsto z^*$. K^* and K are finite sets, but $\hat{K}$ is a union of a finite collection of pairwise disjoint, compact, convex sets.

For $x \in X$ there are defined *carriers* $q^*(x), q(x)$ and $\hat{q}(x)$ in K^*, K and $\hat{K}$. The simplex of K^* whose simplex interior contains x is $q^*(x)$. Similarly, for $q(x)$. $\hat{q}(x)$ is the unique pair $(z^*, t) \in \hat{K}$ such that $z^* = q^*(x)$ and x is in the image, denoted $\langle z^*, t \rangle$, of the local inverse map $\overline{g}_{z^*,t}$.

We use the *inclusion relation* $J = \{(z^*, z) \in K^* \times K : z^* \subset z\}$ to regard $g : K^* \to K$ and $J : K^* \to K$ as an example of the two alphabet case described in Chapter 1 with relations $G^* = J^{-1} \circ g$ on K^* and $G = g \circ J^{-1}$ on K. We define $\hat{G}$ on $\hat{K}$ to be $(\hat{p} \times \hat{p})^{-1}(G^*)$ so that $(z_1^*, t_1) \in \hat{G}(z_0^*, t_0)$ iff $z_1^* \in G^*(z_0^*)$.

The shifts $s_{G^*}^+$ on $K_{G^*}^{*+} \subset K^{*\mathbf{Z}_+}$ and $s_{\hat{G}}^+$ on $\hat{K}_{\hat{G}}^+ \subset \hat{K}^{\mathbf{Z}_+}$ are related to the p.l. map g on X by the functions $q^+ : X \to K^{*+}_{G^*}$ and $\hat{q}^+ : X \to \hat{K}^+_{\hat{G}}$ defined by $q^+(x)_i = q^*(g^i(x))$ and $\hat{q}^+(x)_i = \hat{q}(g^i(x))$ for all $i \in \mathbf{Z}_+$. However, q^+ and $\hat{q}^+$ are not continuous. The central object of our study is a map going the other way.

$$\hat{h}^+ = \{((\mathbf{z}^*, \mathbf{t}), x) \in \hat{K}^+_{\hat{G}} \times X : g^i(x) \in \langle z^*_i, t_i \rangle \text{ for all } i \in \mathbf{Z}_+\}$$

defines a continuous function from $\hat{K}^+_{\hat{G}}$ to X mapping the shift $s^+_{\hat{G}}$ to g. The closed relation $h^+ \circ (\hat{p})^{-1} \subset K^{*+}_{G^*} \times X$ restricts to a function on various important invariant subsets of $K^{*+}_{G^*}$. The proof that $\hat{h}^+$ is a function uses the contraction results of the previous chapter. They also yield a Partial Shadowing Lemma: If $\epsilon > 0$, $(\mathbf{z}^*, \mathbf{t}) \in \hat{K}^+_{\hat{G}}$ and $\xi \in X^{\mathbf{Z}_+}$ are such that $\xi_i \in \langle z^*_i, t_i \rangle$ and $d_K(g(\xi_i), \xi_{i+1}) \leq \epsilon$ for all $i \in \mathbf{Z}_+$ then with $x = \hat{h}^+(\mathbf{z}^*, \mathbf{t})$ $d_K(\xi_i, g^i(x)) \leq C\epsilon$ for all $i \in \mathbf{Z}_+$ where the constant C depends only on the simplicial map g.

Because we can explicitly describe when two elements of $\hat{K}^+_{\hat{G}}$ are mapped to the same point of X by $\hat{h}^+$, we obtain various semiconjugacy results. For example, suppose that K^*_1 is a subdivision of K *isomorphic* to K^*. That is, there is a simplicial isomorphism $r_1 : K^*_1 \to K^*$ such that $r(z) = z$ for all $z \in K$. Then $g_1 = g \circ r_1 : K^*_1 \to K$ is a simplicial dynamical system. There exists a homeomorphism ρ_1 on X such that $\rho_1(z) = z$ for all $z \in K$ and $\rho_1 \circ g_1 = g \circ \rho_1$. That is, the topological dynamical systems associated with the p.l. maps g_1 and g on X are conjugate. However, ρ_1 is *not* the p.l. map associated with r_1. In fact, ρ_1 is usually not p.l. at all.

Chapter 4, Recurrence and Basic Set Images: Let $g : K^* \to K$ be a simplicial dynamical system with $X = |K|$. From the two alphabet results of Chapter 1 we associate to each basic set B^* for the relation $G^* = J^{-1} \circ g$ on K^* a basic set B for $G = g \circ J^{-1}$ on K and vice-versa. The simplices of B^* and of B all have the same dimension. We call B^* and B k *skeleton basic sets* when this common dimension is k.

When B^* is a basic set for G^* then the relation h^+ restricts to a continuous map on $B^{*+}_{G^*} = B^{*\mathbf{Z}_+} \cap K^{*+}_{G^*}$. We call $h^+(B^{*+}_{G^*})$ the associated *basic set image* in X. The restriction of g to each basic set image is a topologically transitive map with dense periodic points. Conversely, if x is a recurrent point for g on X then the sequence of carriers: $q^*(x), q^*(g(x))$,... lies entirely in a single basic set for G^* called the *endset* of x and x lies in the corresponding basic set image. Thus, the union of the basic set images for g is the *Birkhoff center* of g, i.e. the closure of the set of recurrent points. From the basic set images one can construct the–usually larger–basic sets for g and the entire chain recurrent set for g. Each basic set image can be constructed using a procedure mimicking the *iterated function systems* of Barnsley.

Let B^* be a k skeleton basic set for G^* with B the associated basic set for G. It is always true that $h^+(B^{*+}_{G^*}) \subset |B^*| \subset |B|$. If these inclusions are equalities, and so the basic set image is the k dimensional polyhedron $|B|$, then B^* is called a *polyhedral* basic set. Otherwise both inclusions are strict and $h^+(B^{*+}_{G^*})$ is nowhere dense in $|B|$. In that case B^* is called *tattered.* If $h^+(B^{*+}_{G^*})$ meets the interior of some k simplex of B then B^* is called an *interior* basic set. So polyhedral implies interior but there exist interior basic sets which are tattered. The endsets are all interior and every basic set image is contained in some interior basic set image.

The interior basic sets are of special interest because if B^* is interior then the restriction of h^+ to $B^{*+}_{G^*}$ is an *almost homeomorphism* onto its image. That is, for x in a dense G_δ subset of $h^+(B^{*+}_{G^*})$ the set $(h^+)^{-1}(x) \cap B^{*+}_{G^*}$ consists of a single point, namely $q^+(x)$.

When g is a nondegenerate simplicial map, i.e. every $z^* \in K^*$ is nondegenerate and so $\hat{p} : \hat{K} \to K^*$ is a bijection, then there are special results. Suppose X is *everywhere d dimensional*, i.e. $X = |S^d(K^*)|$ where $S^d(K^*) = \{z^* : \dim z^* = d\}$. Then the restriction of h^+ to $S^d(K^*)^+_{G^*} = S^d(K^*)^{\mathbf{Z}_+} \cap K^{*+}_{G^*}$ is an almost homeomorphism onto X mapping the subshift to g.

Chapter 5, Invariant Measures: For g the p.l. map associated with the simplicial dynamical system $g : K^* \to K$, the invariant measures come from the shift. That is, if μ is an ergodic invariant measure for g there exists an interior basic set B^* for G^* and an ergodic invariant measure ν for the shift on $B^{*+}_{G^*}$ such that $\mu = h^+_*\nu$. In fact, h^+ is an isomorphism between the measurable dynamical systems $(s^+_{G^*}, B^{*+}_{G^*}, \nu)$ and (g, X, μ). Of particular importance are the measures which come from Markov measures on the shift.

The relation $G^*_{B^*}$ obtained by restricting the relation G^* to the basic set B^* has a *characteristic matrix* from which is obtained a Markov chain on B^* and an associated Markov measure, ν^P, the *Parry measure* on $B^{*+}_{G^*}$.

Its entropy is $\ln \gamma$ where γ is the dominant eigenvalue of the characteristic matrix for B^*. If B^* is an interior basic set then $\ln \gamma$ is also the topological entropy of g on the basic set image $h^+(B^{*+}_{G^*})$. The topological entropy of g is obtained by letting B^* vary over the interior basic sets and taking the maximum of the $\ln \gamma$'s.

For each simplex z of K with $\dim z = k$ there is a k dimensional Lebesgue measure λ_z on z normalized by $\lambda_z(z) = 1$. If B^* is a polyhedral basic set for G^* with associated G basic set B, there is a unique positive vector p such that the $\sum p_z = 1$ and $\sum p_z \lambda_z = \lambda_B$ (summing over $z \in B$) is an invariant measure for g on $h^+(B^{*+}_{G^*}) = |B|$. The invariant Lebesgue measure λ_B is the projection via h^+ of a Markov measure on $B^{*+}_{G^*}$ and is ergodic. The set of x in X such that $\{g^i(x)\}$ is eventually in some polyhedral basic set image $|B|$ is residual and intersects each $z \in K$ in a subset of λ_z measure 1.

If K_1^* is a subdivision of K isomorphic to K^* by the simplicial isomorphism $r_1 : K_1^* \to K$ and $g_1 = g \circ r_1$: $K_1^* \to K$ is the isomorphic simplicial dynamical system then the relation G on K is the same for g and g_1, and the relations G_1^* on K_1^* and G^* on K^* are related by r_1. If B^* is a polyhedral basic set for G^* then $B_1^* = r_1^{-1}(B^*)$ is a polyhedral basic set for G_1^* with the same associated G basic set B. The Chapter 3 conjugacy homeomorphism ρ_1 on X mapping g_1 to g restricts to a homeomorphism of $|B|$. However, ρ_1 does not relate the Lebesgue measures λ_1 for g_1 and λ for g on $|B|$. In fact, the measurable systems $(g, |B|, \lambda_1)$ and $(g, |B|, \lambda)$ often have different entropy. As they are different measures ergodic for g, λ and $\rho_{1*}\lambda_1$ are, in that case, mutually singular. There exists cases where for no isomorph of g does Lebesgue measure yield the topological entropy.

Chapter 6, Generalized Simplicial Dynamical Systems: Suppose $g : K^* \to K$ is a simplicial map with K^* a subdivision of K but not a proper subdivision. Although g is not a simplicial dynamical system it can nonetheless happen that $\hat{h}^+ : \hat{K}_{\hat{G}} \to X$ is a function and that all of the earlier results are true for these *generalized simplicial dynamical systems.* In this chapter we illustrate how such systems arise and we characterize them.

Chapter 7, Examples: After the higher dimensional Tent Map the examples are all on L = a single 2 simplex together with its faces. We exhibit examples with fractal basic set images and with nonwandering points which are not contained in any basic set image. Also there is an example with entropy zero despite having a basic set shift (but not an interior basic set shift) with positive entropy.

Chapter 8, PL Roundoffs of a Continuous Map: After the details needed for the construction of the p.l. roundoff maps as described above, we consider certain special cases satisfying a general position condition. In these cases the roundoff maps have certain additional structure, a filtration similar to one discussed in Akin, Hurley and Kennedy (1996) Proposition 6.3.

Chapter 9, Nondegenerate Maps on Manifolds: If $g : K^* \to K$ is a simplicial dynamical system and the polyhedron $X = |K|$ is a manifold then there exists a subdivision K_1^* of K^* and a nondegenerate simplicial map $g_1 : K_1^* \to K$ such that the uniform distance between the p.l. maps g_1 and g on X is bounded by 4 times the mesh of K. From this one can prove that for a compact, connected PL manifold X, the set of chain transitive maps on X is a nonempty uniformly closed subset of the space of continuous maps on X and that it contains the set of weak mixing maps as a residual subset.

Appendix: Chapter 10, Stellar and Lunar Subdivisions: We review the simplicial constructions associated with *stellar subdivisions* of a complex and introduce the related idea of a *lunar subdivision.*

Appendix: Chapter 11, Hyperbolicity for Relations: We review the concepts of shadowing and expansivity for homeomorphisms and extend them to closed relations. In particular, we prove that each relation concept is equivalent to the corresponding homeomorphism concept applied to the sample path shift homeomorphism.

* *

Since the magic in all these results is not very subtle, I will violate the magicians' code and reveal the trick. A simplicial dynamical system $g : K^* \to K$ has built in hyperbolicity. The little simplices of K^* are stretched to fit on simplices of K. That is the expanding part. The (super-) contracting part consists of the linear singularities of g on the degenerate simplices. In particular, the nondegenerate cases behave somewhat like expanding maps.

What about more general p.l. maps? Let K be a complex on X and $g : X \to X$ any p.l. map. There exist subdivisions K_1 and K_2 of K such that $g : K_1 \to K_2$ is simplicial. Let K^* be a common subdivision of K_1 and K_2 and for $\alpha = 1, 2$ let $J_\alpha : K^* \to K_\alpha$ be the corresponding inclusion relation. Define the relation G^* on K^* by

$$G^* = J_2^{-1} \circ g \circ J_1.$$

For each point $x \in X$ there exist G^* chains $\mathbf{z}^* \in K^{*\mathbf{Z}_+}$ such that $g^i(x) \in \mathbf{z}_i^*$ for all $i \in \mathbf{Z}_+$. The trouble is that now given $\mathbf{z}^* \in L_{G^*}^{*+}$ such an x need not exist for $\mathbf{z}^*$ and need not be unique when it does exist even in the case when $\mathbf{z}^*$ is periodic.

It might be worth considering such extensions of simplicial dynamical systems because a smooth map on a smooth manifold probably has p.l. approximations which are somewhat better than merely C^0 while the p.l. roundoff maps, degenerate as they are, are inevitably only C^0 close. However, such further considerations are a story for another day.

* *

Thank you to Kate March for preparing the manuscript and to Geoffrey March for the figures. I also want to express my appreciation to Mike Boyle, Fern Hunt, Walter Miller and Dan Rudolph for helpful discussions. I am particularly grateful to my City College colleague Hironari Onishi for the beautiful examples which I included in Chapter 7.

1. Chain Recurrence and Basic Sets

We recall from Akin (1993) some of the notation for and results about relations between compact metric spaces.

A *relation* F from X_1 to X_2, written $F : X_1 \to X_2$, is just a subset of $X_1 \times X_2$. For $A \subset X_1$ the *image* set $F(A) = \{y : (x, y) \in F$ for some $x \in A\}$. So $F(x)$ is the, possibly empty, set of points associated to $x \in X_1$. The relation is a function, or map, when $F(x)$ is a singleton set for every $x \in X_1$. Thus, we identify a function with its graph. For example, the identity map 1_X is the diagonal subset of $X \times X$. A closed relation is just a closed subset.

For a relation $F : X_1 \to X_2$ the *inverse* relation $F^{-1} = \{(y, x) : (x, y) \in F\}$ so that $F^{-1} : X_2 \to X_1$. So for $B \subset X_2$ $F^{-1}(B) = \{x : F(x) \cap B \neq \emptyset\}$. In particular, $F^{-1}(X_2) = \{x : F(x) \neq \emptyset\}$ is called the *domain* of F. The relation is called *surjective* when $F(X_1) = X_2$ and $F^{-1}(X_2) = X_1$, i.e. the domain is all of X_1 and the image is all of X_2. If $G : X_2 \to X_3$ then the composed relation $G \circ F = \{(x, z) : z \in G(y)$ for some $y \in F(x)\}$ so that $G \circ F : X_1 \to X_3$.

When $X_1 = X_2 = X$ we call F a relation or map *on* X. We can then iterate defining $F^0 = 1_X$, $F^1 = F$ and $F^{n+1} = F \circ F^n$ for $n = 1, 2, \ldots$. For a negative n define $F^n = (F^{-1})^{|n|}$. Because composition of relations is associative we have $F^m \circ F^n = F^{m+n}$ provided m and n have the same sign. But $F \circ F^{-1}$ and $F^{-1} \circ F$ need not be the identity map F^0. The *cyclic set* of the relation F is $|F| = \{x : (x, x) \in F\}$, i.e. the domain of $1_X \cap F$. The relation is *reflexive* when $|F| = X$, *symmetric* when $F = F^{-1}$ and *transitive* when $F^2 \subset F$. For a relation F on X the associated *orbit relation* is

$$\mathcal{O}F = \bigcup_{n=1}^{\infty} F^n. \tag{1.1}$$

$\mathcal{O}F$ is the smallest transitive relation containing F. When F is a map $\mathcal{O}F(x)$ is just the positive portion of the orbit sequence of F, regarded as a subset of X.

For a closed relation F on X the associated *chain relation* of Conley (1978) is

$$\mathcal{C}F = \bigcap_{\epsilon > 0} \mathcal{O}(\overline{V}_\epsilon \circ F), \tag{1.2}$$

where $\overline{V}_\epsilon = \{(x, y) : d(x, y) \leq \epsilon\}$. $\mathcal{C}F$ is a closed transitive relation which depends only on the topology of X and not on the choice of metric d (Akin (1993) Proposition 1.8 and Exercise 1.9).

Let $\epsilon \geq 0$. A sequence $\{x_i : i \in I\}$, where I is a subinterval of length at least one in the set of integers $\mathbf{Z}$, is called an ϵ *chain* for F if $x_{i+1} \in \overline{V}_\epsilon(F(x_i))$ whenever $i, i+1 \in I$. In particular, the sequence is a chain, i.e. a 0 chain, if $x_{i+1} \in F(x_i)$ whenever $i, i+1 \in I$. Thus, $y \in \mathcal{O}F(x)$ if there is some chain which begins at x and ends at y, while $y \in \mathcal{C}F(x)$ if for every positive ϵ there is some ϵ chain which begins at x and ends at y. by Proposition 1.11 of Akin (1993)

$$\mathcal{O}(F^{-1}) = (\mathcal{O}F)^{-1}$$

$$\mathcal{C}(F^{-1}) = (\mathcal{C}F)^{-1} \tag{1.3}$$

and so we can omit the parentheses.

The cyclic set of $\mathcal{C}F$, $|\mathcal{C}F|$, is called the *chain recurrent set* for F. Restricted to this set the relation $\mathcal{C}F$ is reflexive as well as transitive and $\mathcal{C}F \cap \mathcal{C}F^{-1}$ is a closed equivalence relation on $|\mathcal{C}F|$. The equivalence classes are called the *basic sets* for the closed relation F. $\mathcal{C}F$ induces a partial order on the space of basic sets (see Akin (1993) Chapter 3). If F is a continuous map and B is a basic set for F then B is F invariant, i.e. $F(B) = B$ (cf. Akin (1993) Corollary 4.11). Such invariance need not hold for more general relations.

Assume that F_α is a closed relation on X_α for $\alpha = 1, 2$. A function, $h : X_1 \to X_2$ *maps* F_1 *to* F_2 when

$$h \times h(F_1) \subset F_2, \tag{1.4}$$

where $h \times h : X_1 \times X_1 \to X_2 \times X_2$ is the product map. Thus, h maps F_1 to F_2 when $y \in F_1(x)$ implies $h(y) \in F_2(h(x))$.

Assume now that h is continuous and satisfies (1.4). Given $\epsilon_2 > 0$, let $\epsilon_1 > 0$ be an ϵ_2 modulus of uniform continuity for h, i.e. $d_1(x, y) \leq \epsilon_1$ implies $d_2(h(x), h(y)) \leq \epsilon_2$. If $\{x_i : i \in I\}$ is an ϵ_1 chain for F_1 then $\{h(x_i) : i \in I\}$ is an ϵ_2 chain for F_2. The same argument with $\epsilon_1 = \epsilon_2 = 0$ shows that h maps an F_1 chain to an F_2 chain. It follows that h maps $\mathcal{C}F_1$ to $\mathcal{C}F_2$, as well, i.e. (1.4) implies

$$h \times h(\mathcal{C}F_1) \subset \mathcal{C}F_2. \tag{1.5}$$

Hence, if $(x,x) \in \mathcal{C}F_1$ then $(h(x), h(x)) \in \mathcal{C}F_2$, i.e.

$$h(|\mathcal{C}F_1|) \subset |\mathcal{C}F_2|. \tag{1.6}$$

In fact, if B_1 is a basic set for F_1 then all of the points of $h(B_1)$ are $\mathcal{C}F_2 \cap \mathcal{C}F_2^{-1}$ equivalent and so there is a unique basic set B_2 for F_2 such that

$$h(B_1) \subset B_2. \tag{1.7}$$

A subset A of X is called $+$ *invariant* with respect to a relation F on X if $F(A) \subset A$, and *invariant* if $F(A) = A$. For example, X itself is always $+$ invariant but need not be invariant when F is not surjective. For any subset A we define the *restriction* of F to A to be the relation

$$F_A = F \cap (A \times A). \tag{1.8}$$

A is called a *surjective* subset when F_A is a surjective relation on A and so when

$$A \subset F(A) \cap F^{-1}(A). \tag{1.9}$$

When A and F are closed then F_A is a closed relation on the space A and the inclusion map from A to X maps F_A to F. So from (1.5) and (1.6) follows

$$\begin{aligned} &\mathcal{C}(F_A) \subset (\mathcal{C}F)_A \\ &|\mathcal{C}(F_A)| \subset |\mathcal{C}F| \cap A. \end{aligned} \tag{1.10}$$

The sample path construction associates maps to a relation in a natural way.

With $\mathbf{Z}$ the set of integers and $\mathbf{Z}_+$ the set of nonnegative integers, we let $X^{\mathbf{Z}}$ and $X^{\mathbf{Z}_+}$ denote the set of biinfinite and infinite sequences in X regarded as functions to X from $\mathbf{Z}$ and $\mathbf{Z}_+$, respectively. With the product topology these are compact, metrizable spaces. For $i \in \mathbf{Z}$ (or $i \in \mathbf{Z}_+$) let $\pi_i : X^{\mathbf{Z}} \to X$ (resp. $\pi_i : X^{\mathbf{Z}_+} \to X$) denote the coordinate i projection. The shift maps $s : X^{\mathbf{Z}} \to X^{\mathbf{Z}}$ and $s : X^{\mathbf{Z}_+} \to X^{\mathbf{Z}_+}$ are a homeomorphism and continuous surjection respectively. That is, for $\xi \in X^{\mathbf{Z}}$ and $i \in \mathbf{Z}$ or $\xi \in X^{\mathbf{Z}_+}$ and $i \in \mathbf{Z}_+$:

$$\pi_i(\xi) = \xi_i$$

$$s(\xi)_i = \xi_{i+1}. \tag{1.11}$$

We denote by $\pi^+ : X^{\mathbf{Z}} \to X^{\mathbf{Z}_+}$ the projection induced by the inclusion of $\mathbf{Z}_+$ into $\mathbf{Z}$.

For a relation F on X we denote by X_F and X_F^+ the *sample path spaces,* the set of chains for F in $X^{\mathbf{Z}}$ and $X^{\mathbf{Z}_+}$, respectively:

$$X_F = \{\xi \in X^{\mathbf{Z}} : \xi_{i+1} \in F(\xi_i) \text{ for all } i \in \mathbf{Z}\}$$

$$X_F^+ = \{\xi \in X^{\mathbf{Z}_+} : \xi_{i+1} \in F(\xi_i) \text{ for all } i \in \mathbf{Z}_+\}. \tag{1.12}$$

X_F is an s invariant subset of $X^{\mathbf{Z}}$ and X_F^+ is an $s+$ invariant subset of $X^{\mathbf{Z}_+}$. When F is closed, these are closed subsets and so we can define the homeomorphism s_F on X_F and the continuous map s_F^+ on X_F^+ by restricting the shift maps. We define $\pi_F^+ : X_F \to X_F^+$ by restricting π^+. Clearly, $\pi_0 : X_F \to X$ maps s_F to F and $\pi_0^+ : X_F^+ \to X$ maps s_F^+ to F. π_F^+ maps s_F to s_F^+ and satisfies $\pi_0^+ \circ \pi_F^+ = \pi_0$.

It is easy to prove that the image of $\pi_0 : X_F \to X$ is the maximum surjective subset of X and satisfies

$$\pi_0(X_F) = \bigcap_{i \in \mathbf{Z}} F^i(X), \tag{1.13}$$

See Miller and Akin (1998) Lemma 1.2, or Proposition 11.1 of the Appendix. Thus, F is a surjective relation iff π_0 is a surjective map on X_F. This subset, called the *dynamic domain* of F, satisfies

$$X \supset \pi_0^+(X_F^+) \supset \pi_0(X_F) \supset |\mathcal{C}F|. \tag{1.14}$$

The first two inclusions are clear and the last is Lemma 4.1 of Miller and Akin (1998).

These constructions are functorial. Also, if $h : X_1 \to X$ is continuous, mapping F_1 to F and F_1 is a map on X_1 then there is a unique lift $\tilde{h}^+ : X_1 \to X_F^+$ such that $\tilde{h}^+$ maps F_1 to s_F^+ and $\pi_0^+ \circ \tilde{h}^+ = h$. If F_1 is a homeomorphism on X_1 then there is a unique lift $\tilde{h} : X_1 \to X_F$ mapping F_1 to s_F with $\pi_0 \circ \tilde{h} = h$. For $y \in X_1$, $\xi = \tilde{h}^+(y)$ or $\xi = \tilde{h}(y)$ is defined by

$$\xi_i = h(F_1^i(y)) \tag{1.15}$$

for all $i \in \mathbf{Z}_+$ or $i \in \mathbf{Z}$, respectively.

For example, if F is itself a map then with $F_1 = F$ and $h = 1_X$ the lift is an inverse showing that $\pi_0^+ : X_F^+ \to X$ is a homeomorphism identifying s_F^+ and F. If F is a homeomorphism then $\pi_0 : X_F \to X$ is a homeomorphism identifying s_F and F.

Similarly, we lift to obtain an identification between s_F on X_F, the sample path homeomorphism for F, and the sample path homeomorphism on $X_{s_F^+}$ for the continuous map s_F^+ on X_F^+. The following diagram of projections and identifications commutes:

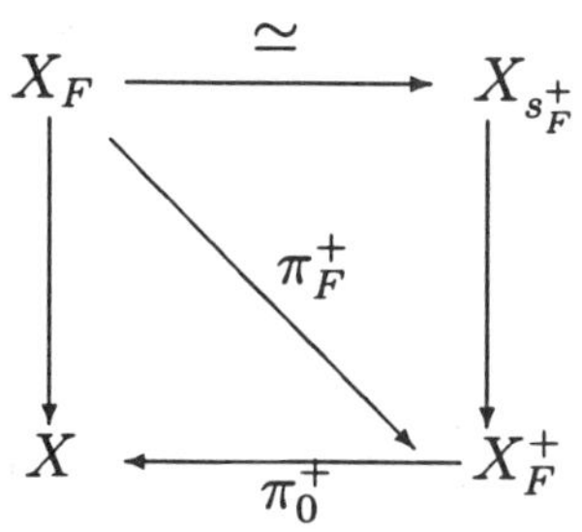

(1.16)

1 Proposition. *Let F be a closed relation on X. For B a basic set for F, define $B_F = X_F \cap B^{\mathbf{Z}}$ and $B_F^+ = X_F^+ \cap B^{\mathbf{Z}_+}$, i.e. the sample spaces for the restriction F_B. B_F is a basic set for s_F mapped by π_0 onto B and is the unique basic set for s_F mapped by π_0 into B. Similarly, B_F^+ is a basic set for s_F^+ mapped by π_0^+ onto B and is the unique basic set for s_F^+ mapped into B. Furthermore, π_F^+ maps B_F onto B_F^+. Thus, $\pi_0 : X_F \to X$, $\pi_0^+ : X_F^+ \to X$ and $\pi_F^+ : X_F \to X_F^+$ induce bijections between the sets of basic sets. It follows that*

$$\pi_0(|\mathcal{C}s_F|) = |\mathcal{C}F| = \pi_0^+(|\mathcal{C}s_F^+|)$$

$$\pi_F^+(|\mathcal{C}s_F|) = |\mathcal{C}s_F^+|. \tag{1.17}$$

Finally, the chain relations satisfy:

$$\pi_0^+ \times \pi_0^+(\mathcal{C}s_F^+) = (\mathcal{C}F) \cap (\pi_0^+(X_F) \times \pi_0^+(X_F))$$

$$\pi_0 \times \pi_0(\mathcal{C}s_F) = (\mathcal{C}F) \cap (\pi_0(X_F) \times \pi_0(X_F)). \tag{1.18}$$

Proof. By Miller and Akin (1998) Lemma 4.1, the basic set B is a surjective subset. From this it easily follows that

$$\pi_0(B_F) = B = \pi_0^+(B_F^+)$$

$$\pi_F^+(B_F) = B_F^+. \tag{1.19}$$

By restricting the relation F to the surjective subset $\pi_0(X_F)$ we obtain the s_F results from Theorem 4.3 and Lemma 4.1 of Miller and Akin (1998). To obtain the s_F^+ results we apply these to the relation s_F^+ and use the identification in diagram (1.16). We then get that each s_F^+ basic set is of the form $\pi_F^+(B_F)$ for some B and so is B_F^+ by (1.19). □

If A is a finite set we regard it as a compact, discrete space. If F is a relation on A then because $\overline{V}_\epsilon = 1_X$ for positive ϵ small enough it follows that $\mathcal{O}F = \mathcal{C}F$. So the basic sets for F are the $\mathcal{O}F \cap \mathcal{O}F^{-1}$ equivalence classes in $|\mathcal{O}F|$.

On $A^{\mathbf{Z}}$ and $A^{\mathbf{Z}_+}$ the homeomorphism s and the map s^+ are the full two-sided shift and the full one-sided shift associated with finite alphabet A. The homeomorphism s_F on A_F and the map s_F^+ on A_F^+ are the *subshifts of finite type* associated with F. The above proposition implies that the basic sets for s_F and s_F^+ are all of the form B_F and B_F^+ for a (uniquely defined) basic set B for F on A.

For such an F basic set B the associated basic sets B_F and B_F^+ of s_F and s_F^+ are just the sample path spaces for the restricted relation F_B. The homeomorphism s_F on B_F and the map s_F^+ on B_F^+ are topologically transitive with dense periodic points (see e.g. Appendix: Chapter 11). In addition, if the restriction F_B is not periodic, that is it does not map onto a nontrivial cycle, then the shift maps are topologically mixing.

If f is a continuous map on X then f is called *topologically transitive* if for some x the positive orbit set $\mathcal{O}f(x)$ is dense in X, in which case the set of such points, the *transitive points* for f, is residual in X, see e.g. Akin (1993) Theorem 4.12. Also, f is topologically transitive iff for every pair of nonempty open subsets U and V of X, $U \cap f^{-i}(V) \neq \emptyset$ for infinitely many positive integers i. If for all such U and V the sets $U \cap f^{-i}(V)$ are nonempty for all sufficiently large i then f is called *topologically mixing.* A *periodic point* x for f is a point of $|\mathcal{O}f|$ i.e. there exists a positive integer m such that $f^m(x) = x$. If f is topologically transitive and $|\mathcal{O}f|$ is dense in X then for every $\epsilon > 0$ there exists a periodic point whose orbit is ϵ dense, i.e. for some $m > 0$, $f^m(x) = x$ and $\cup_{i=0}^{m-1}\overline{V}_\epsilon(f^i(x)) = X$. This claim is a special case of the following result of John Taylor.

2 Lemma. *Let f be a topologically transitive continuous map on X and let*

D be a dense subset of X. For every $\epsilon > 0$, there exists $x \in D$ and $m > 0$ such that $\{x, f(x), \ldots, f^m(x)\}$ is ϵ dense in X.

Proof. If $\mathcal{O}f(\tilde{x})$ is dense in X, then for some $m > 0$ $\{\tilde{x}, f(\tilde{x}), \ldots, f^m(\tilde{x})\}$ is $\epsilon/2$ dense in X. Let $\delta > 0$ be an $\epsilon/2$ modulus of uniform continuity for $f, \ldots, f^m$. Any $x \in D \cap \overline{V}_\delta(\tilde{x})$ will suffice. □

In particular, this result applies to s_F on B_F and s_F^+ on B_F^+ for B a basic set for the relation F on the finite set A.

For the relation F on A itself the following will prove useful.

3 Lemma. *Let $\{x_i : i = 0, 1, \ldots\}$ be an F chain in A indexed by $\mathbf{Z}_+$. There is an F basic set B and an index $i_0 \in \mathbf{Z}_+$ such that $x_i \in B$ for all $i \geq i_0$.*

Proof. If $x_{i_0} = x_{i_1}$ with $i_0 < i_1$ then this element of A is in $|\mathcal{O}F|$ and is $\mathcal{O}F \cap \mathcal{O}F^{-1}$ equivalent to each x_j with j between i_0 and i_1. For some $a \in A$, $x_i = a$ for infinitely many $i \in \mathbf{Z}_+$. Let i_0 be the smallest index such that $x_{i_0} = a$. For every $j > i_0$ there exists $i_1 > j$ such that $x_{i_1} = a$. Hence, x_j is $\mathcal{O}F \cap \mathcal{O}F^{-1}$ equivalent to a. □

The finite set relations we will consider arise in a paired fashion. Consider two finite sets A^* and A, thought of as large and small, respectively. We are given a map $g : A^* \to A$ and a relation $J : A^* \to A$. We then obtain two relations:

$$G = g \circ J^{-1} : A \to A$$

$$G^* = J^{-1} \circ g : A^* \to A^*. \tag{1.20}$$

Because $g \circ G^* = G \circ g$ it follows that g maps G^* to G. Writing g also for the product of copies of g on $A^{*\mathbf{Z}}$ and $A^{*\mathbf{Z}_+}$ we have that g maps s_{G^*} on $A^*_{G^*}$ to s_G on A_G and $s^+_{G^*}$ on $A^{*+}_{G^*}$ to s^+_G on A^+_G. It follows that g maps G^* basic sets to G basic sets, but more is true in this case.

4 Proposition. *For finite sets A^*, A assume $g : A^* \to A$ is a map and $J : A^* \to A$ is a relation. Let $G^* = J^{-1} \circ g$ and $G = g \circ J^{-1}$ be the associated relations on A^* and A, respectively.*

The map g induces a bijection from the set of G^* basic sets to the set of G basic sets. If B^* is a G^* basic set then $B = g(B^*)$ is a G basic set called the associated G basic set for B^*. If B is a G basic set then

$B^* = g^{-1}(B) \cap J^{-1}(B)$ is a G^* basic set, the unique G^* basic set to which B is associated.

The maps $g : A^*_{G^*} \to A_G$ and $g : A^{*+}_{G^*} \to A^+_G$ induce bijections between the sets of basic sets. If B^* is a G^* basic set and B is the associated G basic set then

$$(1.21) \qquad g(B^*_{G^*}) = B_G \quad \text{and} \quad g(B^{*+}_{G^*}) = B^+_G.$$

Proof. By relabeling if necessary we can assume that A^* and A are disjoint sets. Let $\tilde{A} = A^* \cup A$ and define

$$(1.22) \qquad \tilde{G} = g \cup J^{-1} \subset (A^* \times A) \cup (A \times A^*) \subset \tilde{A} \times \tilde{A}.$$

$\tilde{G}$ is a relation with $\tilde{G}^2 = G^* \cup G$. The relation maps onto the two cycle: $A^* \to A \to A^*$. From this it is clear that

$$(\mathcal{O}\tilde{G}) \cap (A^* \times A^*) = \mathcal{O}G^*$$
$$(1.23) \qquad (\mathcal{O}\tilde{G}) \cap (A \times A) = \mathcal{O}G.$$

Intersecting with the diagonal $1_{\tilde{A}}$ we see that

$$(1.24) \qquad |\mathcal{O}\tilde{G}| = |\mathcal{O}G^*| \cup |\mathcal{O}G|.$$

For any basic set $\tilde{B}$ for $\tilde{G}$ define

$$(1.25) \qquad B^* = \tilde{B} \cap A^* \quad \text{and} \quad B = \tilde{B} \cap A.$$

By (1.23) two points of A^* are $\mathcal{O}G^* \cap \mathcal{O}G^{*-1}$ equivalent iff they are $\mathcal{O}\tilde{G} \cap \mathcal{O}\tilde{G}^{-1}$ equivalent. Hence, B^* is a G^* basic set and similarly B is a G basic set. On the other hand, if B^*_1 is a G^* basic set then it is contained in a unique $\tilde{B}$ a $\tilde{G}$ basic set and so $B^*_1 \subset \tilde{B} \cap A^* = B^*$. As distinct basic sets are disjoint $B^*_1 = B^*$. Similarly, every G basic set B is of the form $\tilde{B} \cap A$ for a unique $\tilde{G}$ basic set $\tilde{B}$. Now we show that for a $\tilde{G}$ basic set $\tilde{B}$, (1.25) implies

$$(1.26) \qquad B = g(B^*) \quad \text{and} \quad B^* = g^{-1}(B) \cap J^{-1}(B).$$

Because $\tilde{B}$ is a $\tilde{G}$ basic set it is surjective, that is,

$$(1.27) \qquad \tilde{B} \subset \tilde{G}(\tilde{B}) \cap \tilde{G}^{-1}(\tilde{B}).$$

In particular, $B \subset \tilde{G}(\tilde{B}) \cap A = \tilde{G}(\tilde{B} \cap A^*) = \tilde{G}(B^*) = g(B^*)$. On the other hand, g maps B^* into some G basic set and so $g(B^*) = B$. This implies $B^* \subset g^{-1}(B)$. Intersecting with A^* instead, (1.27) yields $B^* \subset \tilde{G}(\tilde{B}) \cap A^* = \tilde{G}(\tilde{B} \cap A) = \tilde{G}(B) = J^{-1}(B)$. On the other hand, if $a^* \in A^*$ with $a_0 \in J(a^*) \cap B$ and $a_1 = g(a^*) \in B$ then $\{a_0, a^*, a_1\}$ is a $\tilde{G}$ chain. As $a_0, a_1 \in B$, a subset of the $\mathcal{O}\tilde{G} \cap \mathcal{O}\tilde{G}^{-1}$ equivalence class $\tilde{B}$, it follows that $a^* \in \tilde{B}$ and so $a^* \in \tilde{B} \cap A^* = B^*$.

If $\{a_0^*, \dots, a_m^*\}$ is a G^* chain then, letting $a_{i+1} = g(a_i^*)$ for $i = 0, \dots, m$, the definition of G^* implies $a_i^* \in J^{-1}(a_i)$ for $i = 1, \dots, m$. Thus, $\{a_0^*, a_1, a_1^*, \dots, a_m, a_m^*, a_{m+1}\}$ is a $\tilde{G}$ chain and $\{a_1, \dots, a_{m+1}\}$ is a G chain. If $a_0^*, a_m^* \in B^*$ and so the original G^* chain lies in B^* then the extended $\tilde{G}$ chain lies in $\tilde{B}$ and the G chain lies in B. On the other hand, if $\{a_0, \dots, a_m\}$ is a G chain then for $i = 0, \dots, m-1$ there exists $a_i^* \in A^*$ such that $a_i^* \in J^{-1}(a_i)$ and $g(a_i^*) = a_{i+1}$. Thus, $\{a_0, a_0^*, a_1, a_1^*, \dots, a_{m-1}^*, a_m\}$ is a $\tilde{G}$ chain and $\{a_0^*, \dots, a_{m-1}^*\}$ is a G^* chain. If $a_0, a_m \in B$ and so the original G chain lies in B then the $\tilde{G}$ chain lies in $\tilde{B}$ and the G^* chain lies in B^*. From these results applied to infinite chains follows (1.21). The bijections between the sets of shift map basic sets are easily established using Proposition 1. □

The classic example of this *two alphabet* picture is a directed graph with vertex set A and edge set A^*. To each edge a^* there is associated an initial vertex $J(a^*)$ and a terminal vertex $g(a^*)$. A pair of vertices (a_1, a_2) is in the vertex relation G iff there exists an edge a^* beginning at a_1 and terminating at a_2. A pair of edges (a_1^*, a_2^*) is in the edge relation G^* if terminus of a_1^* is the initial vertex of a_2^*. Conversely, any two alphabet model where J, as well as g, is a function can be described by such a directed graph.

In our application of Proposition 4 we will have, in addition to the finite sets A^* and A, a compact metric space $\hat{A}$ and a continuous surjection $\hat{p} : \hat{A} \to A$. Then we define the closed relation $\hat{G} = (\hat{p} \times \hat{p})^{-1} G^*$ on $\hat{A}$. The relation $\hat{G}$ and its shift maps $s_{\hat{G}}$ on $\hat{A}_{\hat{G}}$ and $s_{\hat{G}}^+$ on $\hat{A}_{\hat{G}}^+$ will be called *fat shifts* associated with G^*. The map $\hat{p}$ induces surjective continuous maps $\hat{p} : A_{\hat{G}} \to A_{G^*}^*$ and $\hat{p}^+ : A_{\hat{G}}^+ \to A_{G^*}^{*+}$.

Thus, for a fat shift each point a^* of A^* is fattened up to a nonempty compact subset $\hat{p}^{-1}(a^*)$ in $\hat{A}$. We call a^* *nondegenerate* if $\hat{p}^{-1}(a^*)$ is a singleton set. We say that the fat shift has *finite degeneracies* if the points of $|\mathcal{O}G^*|$ are all nondegenerate.

When f is a continuous map on X, we will also use Proposition 1 to apply homeomorphism results. For a continuous map f the dynamic domain $\pi_0(X_f) = \cap_{i=0}^{\infty} f^i(X)$ and so $\pi_0(X_f)$ is the maximum f invariant set in X, i.e. it contains every subset B such that $f(B) = B$. If $x \in X$ then $\omega f(x)$ denotes the limit point set of the positive orbit sequence $\{f^i(x) : i = 1, 2, \ldots\}$. It is a closed invariant subset of X and any two points in $\omega f(x)$ are $\mathcal{C}f \cap \mathcal{C}f^{-1}$ equivalent, (cf. Akin (1993) Proposition 1.12.) So each $\omega f(x)$ is entirely contained in some basic set B_+. When f is a homeomorphism on X we define $\alpha f(x) = \omega(f^{-1})(x)$, the limit point set of the negative orbit sequence. $\alpha f(x)$ is then entirely contained in some basic set B_- as well. The basic sets B_+ and B_- are equal if and only if $x \in |\mathcal{C}f|$ in which case the closure of the entire orbit of x lies in $B_+ = B_-$.

If f_α is a continuous map on X_α $(\alpha = 1, 2)$ and $h : X_1 \to X_2$ is continuous, mapping f_1 to f_2 then for $x \in X_1$:

$$h(\omega f_1(x)) = \omega f_2(h(x)). \tag{1.28}$$

If f_1 and f_2 are both homeomorphisms then h similarly maps $\alpha f_1(x)$ onto $\alpha f_2(h(x))$.

Now assume f is a homeomorphism on X. Assume that $\mathcal{F}$ is a finite, pairwise disjoint collection of closed, nonempty, f invariant subsets of X. $\mathcal{F}$ is called an *invariant decomposition* for f if it covers the set of limit points. This means that for every $x \in X$ there exist $F_-(x), F_+(x) \in \mathcal{F}$ such that

$$\alpha f(x) \subset F_-(x), \ \omega f(x) \subset F_+(x). \tag{1.29}$$

The set of pairs $(F_-(x), F_+(x))$ in $\mathcal{F} \times \mathcal{F}$ is denoted $\mathcal{F}f$. Thus, $F_2 \in \mathcal{F}f(F_1)$ if there exists $x \in X$ such that $\alpha f(x) \subset F_1$ and $\omega f(x) \subset F_2$. For any $F \in \mathcal{F}$ there exists $x \in F$ because F is nonempty. For such x, $\alpha f(x) \cup \omega f(x) \subset F$ because F is closed and invariant. Thus, the relation $\mathcal{F}f$ on the finite set $\mathcal{F}$ is reflexive, i.e. $|\mathcal{F}f| = \mathcal{F}$. For each $F \in \mathcal{F}$ the associated $\mathcal{F}$ *basic set* is defined to be

$$B(F) = \{x : (F_+(x), F), (F, F_-(x)) \in \mathcal{O}(\mathcal{F}f)\}. \tag{1.30}$$

Thus, $x \in B(F)$ for some $F \in \mathcal{F}$ if there is a finite sequence $\{x_0, \ldots, x_n\}$ in X with $x_0 = x_n = x$ and such that $\omega f(x_i)$ and $\alpha f(x_{i+1})$ are contained in the same element of $\mathcal{F}$ for $i = 0, \ldots, n-1$. $F \subset B(F)$ for all $F \in \mathcal{F}$.

Finally, a decomposition $\tilde{\mathcal{F}}$ is called a *fine decomposition* if for each

$F \in \mathcal{F}$ any two points of F are $\mathcal{C}f \cap \mathcal{C}f^{-1}$ equivalent. It suffices to consider only the limit points of F for if $\alpha f(x)$ and $\omega f(x)$ are contained in the same basic set then this basic set contains x as well. Clearly, $\mathcal{F}$ is a fine decomposition exactly when:

$$\bigcup_{F \in \mathcal{F}} F \times F \subset \mathcal{C}f. \tag{1.31}$$

5 Proposition. *Assume $\mathcal{F}$ is an invariant decomposition for a homeomorphism f on X.*

The collection $\{B(F) : F \in \mathcal{F}\} \equiv \mathcal{F}_b$ is an invariant decomposition for f. Every f basic set is contained in a unique $\mathcal{F}$ basic set. In particular,

$$|\mathcal{C}f| \subset \cup\{B(F) : F \in \mathcal{F}\}. \tag{1.32}$$

The decomposition $\mathcal{F}$ is fine iff each $B(F)$ is an f basic set in which case $\mathcal{F}_b$ is exactly the collection of all basic sets for f. In particular, if f admits a fine decomposition then it has only finitely many basic sets. Conversely, if f has only finitely many basic sets then the collection of all the basic sets of f is a fine decomposition.

Proof. These results are portions of Akin (1993) Theorem 5.15 and Proposition 5.17. □

A finite collection $\tilde{\mathcal{F}}$ of closed, nonempty, f invariant subsets of X is called an *invariant pre-decomposition* for a homeomorphism f if for every $x \in X$ there exist $F_-(x)$, $F_+(x)$ in $\tilde{F}$ containing $\alpha f(x)$ and $\omega f(x)$, respectively, i.e. (1.29) holds. That is, we drop the assumption of pairwise disjointness from the definition of a decomposition.

Define on $\tilde{\mathcal{F}}$ the *intersection relation* $\tilde{I}$ to be the set of pairs (F_1, F_2) such that $F_1 \cap F_2 \neq \emptyset$. Thus, $\tilde{I}$ is reflexive and symmetric. It is $1_{\tilde{\mathcal{F}}}$ exactly when $\tilde{\mathcal{F}}$ is a decomposition. By taking the union of each $\mathcal{O}(\tilde{I})$ equivalence class we obtain the *concatenation* $\mathcal{F}$ for $\tilde{\mathcal{F}}$:

$$[F] = \cup\{F_1 : F_1 \in \mathcal{O}\tilde{I}(F)\}$$

$$\mathcal{F} = \{[F] : F \in \tilde{F}\}. \tag{1.33}$$

$\mathcal{F}$ is called *fine* when each $F \in \mathcal{F}$ is contained in some basic set for f.

6 Proposition. *Assume $\tilde{\mathcal{F}}$ is an invariant pre-decomposition for a homeomorphism f on X.*

(a) The concatenation $\mathcal{F}$ of $\tilde{\mathcal{F}}$ is an invariant decomposition for f. Furthermore, $\mathcal{F}$ is a fine decomposition iff $\tilde{\mathcal{F}}$ is a fine pre-decomposition. In particular, f admits a fine pre-decomposition iff it has only finitely many basic sets.

(b) Assume that f_1 is a homeomorphism on X_1 and that $h : X \to X_1$ is surjective continuous map, mapping f to f_1. Define

$$h\tilde{\mathcal{F}} = \{h(F) : F \in \tilde{\mathcal{F}}\}. \tag{1.34}$$

$h\tilde{\mathcal{F}}$ is an invariant pre-decomposition for f_1. Furthermore, $h\tilde{\mathcal{F}}$ is fine if $\tilde{\mathcal{F}}$ is. In particular, if f has only finitely many basic sets, then f_1 has only finitely many basic sets.

Proof. (a) Each $[F]$ is a closed, invariant set containing F and so $\mathcal{F}$ is a pre-decomposition. Clearly, distinct elements of $\mathcal{F}$ are disjoint by definition of $\mathcal{O}\tilde{I}$. So $\mathcal{F}$ is a decomposition. $[F] \subset B$ implies $F \subset B$ and so $\tilde{\mathcal{F}}$ is fine when $\mathcal{F}$ is. If $\tilde{\mathcal{F}}$ is fine and F, $F_1 \in \tilde{\mathcal{F}}$ are contained in basic sets B, B_1 then $F \cap F_1 \neq \emptyset$ implies $B = B_1$ because distinct basic sets are disjoint. As each $F_1 \in \mathcal{O}\tilde{I}(F)$ is contained in B it follows that $[F] \subset B$.

(b) Each $h(F)$ is closed, nonempty and f_1 invariant because h maps f to f_1. For any $y \in X_1$ there exists $x \in X$ such that $h(x) = y$. Then by (1.28) and (1.29), $\omega f_1(y) \subset h(F_+(x))$, and similarly $\alpha f_1(y) \subset h(F_-(x))$. Thus, $h\tilde{\mathcal{F}}$ is an invariant pre-decomposition. If $\tilde{\mathcal{F}}$ is fine then $h\tilde{\mathcal{F}}$ is fine by (1.7).

The association between fine pre-decompositions and finitely many basic sets follows from Proposition 5.

Remark. Observe that if x is a recurrent point, i.e. $x \in \omega f(x)$ then x together with its entire orbit closure is contained in $F_+(x) \in \tilde{\mathcal{F}}$. For example, every periodic orbit is contained in some element of $\tilde{\mathcal{F}}$. □

We conclude this section with some results about continuous maps which are injective on various large subsets.

For a continuous map $h : X_1 \to X_2$ and a subset B of X_2 we say that h is *injective over* B if the restriction of h to $h^{-1}(B)$ is injective, i.e. $y \in B$ implies $h^{-1}(y)$ has at most one element in X_1. Now for $\epsilon > 0$ define

$$\mathrm{Inj}_h^\epsilon = \{y : h^{-1}(y) \times h^{-1}(y) \subset V_\epsilon\} \tag{1.35}$$

where $V_\epsilon = \{(x_1, x_2) : d(x_1, x_2) < \epsilon\}$. So $y \in \mathrm{Inj}_h^\epsilon$ iff the diameter of $h^{-1}(y)$ is less than ϵ. By compactness, there is then some neighborhood U of y such that $h^{-1}(U) \times h^{-1}(U) \subset V_\epsilon$. It follows that Inj_h^ϵ is an open set. Consequently,

$$\text{(1.36)} \qquad \mathrm{Inj}_h = \cap_{\epsilon>0} \mathrm{Inj}_h^\epsilon = \{y : h^{-1}(y) \times h^{-1}(y) \subset 1_{X_1}\}$$

is a G_δ. It is the largest subset of X_2 over which h is injective. A continuous map h is called *almost injective* if $h^{-1}(\mathrm{Inj}_h)$ is a dense, and hence residual, subset of X_1. An almost injective map is called an *almost homeomorphism* when it is surjective as well. In that case, $\mathrm{Inj}_h = hh^{-1}(\mathrm{Inj}_h)$ is a residual subset of X_1 (see Akin (1997) Appendix).

Now assume that f_α is a continuous map on X_α ($\alpha = 1, 2$) and $h : X_1 \to X_2$ is a continuous surjection mapping f_1 to f_2. h is called an *asymptotic extension map* if

$$\text{(1.37)} \qquad |\mathcal{C}f_1| \subset h^{-1}(\mathrm{Inj}_h),$$

i.e. if h is injective over $h(|\mathcal{C}f_1|)$. The name comes from the following result.

7 Proposition. *Assume f_α is a continuous map on X_α ($\alpha = 1, 2$) and $h : X_1 \to X_2$ is an asymptotic extension map. Let $\epsilon > 0$. For every $y \in X_2$ there exists a positive integer N such that $f_2^i(y) \in \mathrm{Inj}_h^\epsilon$ for all $i \geq N$. If $h(x) = h(\tilde{x}) = y$ then for all $i \geq N$*

$$\text{(1.38)} \qquad d(f_1^i(x), f_1^i(\tilde{x})) < \epsilon.$$

If f_1 and f_2 are homeomorphisms then we can choose N so that $f_2^i(y) \in \mathrm{Inj}_h^\epsilon$ for all $i \in \mathbf{Z}$ with $|i| \geq N$ and so (1.38) holds whenever $|i| \geq N$.

Proof. By (1.28), $\omega f_2(y) \subset h(|\mathcal{C}f_1|)$. Because (1.37) is equivalent to

$$\text{(1.39)} \qquad h(|\mathcal{C}f_1|) \subset \mathrm{Inj}_h,$$

Inj_h^ϵ is a neighborhood of the compact set $\omega f_2(y)$. By definition of the ω limit set the sequence $\{f_2^i(y)\}$ is eventually contained in Inj_h^ϵ. If $h(x) = y = h(\tilde{x})$ then $f_1^i(x)$ and $f_1^i(\tilde{x})$ are both in $h^{-1}(f_2^i(y))$. So (1.38) follows from the definition (1.35). In the homeomorphism case we apply the same result to f_α^{-1} on X_α ($\alpha = 1, 2$). □

Suppose that A^* is a finite set with relation G^* and $\hat{p} : \hat{A} \to A^*$ is a continuous surjection inducing $\hat{G} = (\hat{p} \times \hat{p})^{-1}G^*$ on $\hat{A}$. The maps from the fat shifts to the associated subshifts are $\hat{p} : \hat{A}_{\hat{G}} \to A^*_{G^*}$ and $\hat{p}^+ : \hat{A}^+_{\hat{G}} \to A^{*+}_{G^*}$. If $\hat{G}$ has finite degeneracies then it is easy to prove that $\hat{p}$ and $\hat{p}^+$ are asymptotic extension maps.

2. Simplicial Maps and Their Local Inverses

A nonempty, finite subset V of a real vector space is called *independent* if $\sum c_v v = 0$ and $\sum c_v = 0$ (sums over $v \in V$) only when the coefficients $c_v = 0$ for all $v \in V$. When V is independent the convex hull, $[V]$, is called the *simplex* with *vertex set* V, or *spanned by* V. The vertices are the extreme points of the simplex and so the vertex set is determined by the simplex. For a simplex z we let $V(z)$ denote the vertex set of z. The *dimension* of z, denoted dim z, is 1 less than the cardinality of $V(z)$.

For example, for any finite set, V, we let $\mathbf{R}^V$ denote the vector space of real functions on V and for $v \in V$ let e^v be the standard basis vector with $e^v_v = 1$ and $e^v_w = 0$ for $w \in V\backslash\{v\}$, $\{e^v : v \in V\}$ is independent and spans the *unit simplex* Δ_V:

$$\Delta_V = \{c \in \mathbf{R}^V : c_v \geq 0 \text{ for all } v \in V \text{ and } \sum_{v \in V} c_v = 1\}. \tag{2.1}$$

The affine subspace generated by Δ_V is parallel to the linear subspace

$$\mathbf{R}^V = \{c \in \mathbf{R}^V_0 : \sum_{v \in V} c_v = 0\}. \tag{2.2}$$

Observe that dim Δ_V is the vector space dimension of $\mathbf{R}^V_0$. Clearly,

$$a, b \in \Delta_V \Rightarrow a - b \in \mathbf{R}^V_0. \tag{2.3}$$

In general, if z is a simplex with vertex set $V = V(z)$ then for each $x \in z$ the *barycentric coordinate vector* $c_z(x)$ is uniquely defined by:

$$c_z(x) \in \Delta_V \quad \text{and} \quad x = \sum_{v \in V} c_z(x)_v v. \tag{2.4}$$

Thus, the map $c_z : z \to \Delta_V$ is the linear extension of the association $v \mapsto e^v$ for $v \in V(z)$.

If V_1 is a nonempty subset of the vertex set $V(z)$ then V_1 spans a simplex z_1 called a *face* of z, written $z_1 < z$. For example, the vertices are the zero dimensional faces of z and so $v < z$ for a vector v iff $v \in V(z)$. Clearly,

$$z_1 < z \Rightarrow \dim z_1 \leq \dim z$$

$$\text{with equality iff } z_1 = z. \tag{2.5}$$

If $z_1 < z$ and $z_1 \neq z$ then z_1 is called a *proper face* of z. The union, ∂z, of the proper faces is called he *boundary* of z. The complementary set, $z \backslash \partial z$, is called the *interior* of z and is denoted $(z)^\circ$. We leave the following to the reader:

1 Lemma. *Let z be a simplex, $z_1 < z$ and $x \in z$.*

(a) $x \in z_1$ iff $c_z(x)_v = 0$ for all $v \in V(z) \backslash V(z_1)$. In that case. $c_z(x)_v = c_{z_1}(x)_v$ for all $v \in V(z_1)$.

(b) $x \in (z)^\circ$ iff $c_z(x)_v > 0$ for all $v \in V(z)$. □

A *simplicial complex* K is a finite set of simplices in a vector space satisfying the two conditions:

(2.6) $$z \in K \text{ and } z_1 < z \Rightarrow z_1 \in K.$$

(2.7) $$z_1, z_2 \in K \text{ and } z_1 \cap z_2 \neq \emptyset \Rightarrow$$ $$z_1 \cap z_2 \text{ is a common face of } z_1 \text{and } z_2.$$

For any subset L of K we define the subsets of L:

(2.8) $$S^k(L) = \{z \in L : \dim z = k\}$$ $$\overline{S}^k(L) = \{z \in L : \dim z \leq k\}.$$

L is called a *subcomplex* of K if $z \in L$ and $z_1 < z$ imply $z_1 \in L$ and so L itself is a simplicial complex. In that case, each $\overline{S}^k(L)$ is a subcomplex called the k *skeleton* of L. In particular, $S^0(K) = \overline{S}^0(K)$ is the set of vertices of K, also denoted $V(K)$, so that

(2.9) $$V(K) = \cup\{V(z) : z \in K\}.$$

For $L \subset K$ we define

(2.10) $$|L| = \cup\{z : z \in L\},$$

a subset of the ambient vector space. Clearly, if L_1 is the subcomplex consisting of the simplices of L together with their faces then $|L| = |L_1|$. A subset X of a vector space is called a *polyhedron* if $X = |K|$ for some simplicial complex K in which case we say that K triangulates X. Thus, any subcomplex L triangulates $|L|$.

We combine the functions $\{c_z : z \in K\}$ to define the *barycentric coordinate map* for K:

$$c_K : |K| \to \Delta_{V(K)}$$

$$c_K(x)_v = \begin{cases} c_z(x)_v & \text{if } x \in z \text{ and } v \in V(z) \\ 0 & \text{otherwise} \end{cases} \tag{2.11}$$

Thus, $c_K(x)$ extends $c_z(x)$ by assigning 0 as the barycentric coordinates for the remaining vertices. The function is well-defined by Lemma 1a. By (2.4) we have for all $x \in |K|$:

$$x = \sum_{v \in V(K)} c_K(x)_v v. \tag{2.12}$$

If we define for $z \in K$ the *open star* of z:

$$St(z) = \cup\{(z_1)^\circ : z < z_1\} \tag{2.13}$$

then we have for $z \in K$ and $x \in |K|$:

$$x \in z \quad \Leftrightarrow \quad c_K(x)_v > 0 \text{ only if } v \in V(z).$$

$$x \in St(z) \quad \Leftrightarrow \quad c_K(x)_v > 0 \text{ if } v \in V(z).$$

$$x \in (z)^\circ \quad \Leftrightarrow \quad c_K(x)_v > 0 \text{ iff } v \in V(z). \tag{2.14}$$

By (2.12), c_K is injective. A polyhedron has a natural compact topology obtained from the linear structure and c_K is an embedding, i.e. a homeomorpism onto its image. The coordinate map induces a natural metric by using the l^1 norm, $\| c \| = \sum_v |c_v|$ on vectors $c \in \mathbf{R}^V$. For $x_1, x_2 \in |K|$ define

$$\begin{aligned} d_K(x_1, x_2) &= \| c_K(x_1) - c_K(x_2) \| \\ &= \sum_{v \in V(K)} |c_K(x_1)_v - c_K(x_2)_v|. \end{aligned} \tag{2.15}$$

Clearly, $d_K(x_1, x_2) \leq 2$ with equality if x_1 and x_2 lie in disjoint closed simplices so that $c_K(x_1)_v c_K(x_2)_v = 0$ for all $v \in V(K)$. In particular, $d_K(v_1, v_2) = 2$ for any pair of distinct vertices in K.

Each simplex z in K is a compact subset of $|K|$ and by (2.14) $St(z)$ is an open subset of $|K|$. As $(z)^\circ$ is dense in z we have for z_1, $z \in K$:

$$z_1 < z \Leftrightarrow (z_1)^\circ \cap z \neq \emptyset \Leftrightarrow z \cap St(z_1) \neq \emptyset. \tag{2.16}$$

For a simplex z in K we will use the same symbol to denote the element of K, the subset of $|K|$ and the subcomplex of K consisting of z and its faces. In particular, we allow context to select among these meanings for Δ_V.

If A is a nonempty subset of $|K|$ and A is contained in some simplex of K then it is contained in a smallest such. The intersection of $\{z \in K : A \subset z\}$ is a common face of the simplices in this set. This smallest simplex is called the *carrier* (in K) of A. For a point $x \in |K|$ the carrier is the unique simplex z such that $x \in (z)^\circ$. We define q, $Q \subset |K| \times K$:

$$q = \{(x, z) : x \in (z)^\circ\}$$

$$Q = \{(x, z) : x \in z\}. \tag{2.17}$$

The carrier mapping q is usually not continuous. The relation Q is its closure and satisfies (see (2.5)):

$$z \in Q(x) \Leftrightarrow q(x) < z, \text{ in which case}$$

$$\dim q(x) = \dim\ z \quad \text{iff } q(x) = z. \tag{2.18}$$

If K^* and K are simplicial complexes then a *simplicial map* $g : K^* \to K$ is a function satisfying

$$v \in V(K^*) \Rightarrow g(v) \in V(K); \tag{2.19}$$

$$g(V(z^*)) = V(g(z^*)) \text{ for all } z^* \in K^*. \tag{2.20}$$

That is, g takes vertices to vertices and the vertex map determines g in that the vertex set of $g(z^*)$ is just the set of images of the vertices of z^*.

The simplicial map itself is a map between finite sets. We obtain the associated *piecewise linear* (written p.l.) map $g : |K^*| \to |K|$ by extending linearly over each simplex. Using the coordinate function we have for $x \in |K^*|$

$$g(x) = \sum_{v \in V(K^*)} c_{K^*}(x)_v g(v). \tag{2.21}$$

For example, for a simplicial complex K, the association $v \mapsto e^v$ extends to define a simplicial map $c_K : K \to \Delta_{V(K)}$ and the associated p.l. map on $|K|$ is just the barycentric coordinate map itself.

We use the same symbol g for the simplicial map and its p.l. extension because the image of the subset z^* under the p.l. map g is just the subset $g(z^*)$, the value of the simplicial map at z^*.

We say the simplicial map g is *nondegenerate* on the simplex z^* of K^* (or just z^* is a nondegenerate simplex when g is understood) if g is injective on the vertex set of z^*. Clearly,

$$\dim z^* \geq \dim g(z^*)$$

(2.22) with equality iff z^* is nondegenerate.

In particular, we have that

$$g(\overline{S}^k(K^*)) \subset \overline{S}^k(K). \tag{2.23}$$

If g is nondegenerate on z^* then the p.l. map g restricts to a bijective mapping of z^* to $g(z^*)$ and so we can define $\overline{g}_{z^*}$ to be the inverse of the restriction $g|z^* : z^* \to g(z^*)$. That is,

$$\overline{g}_{z^*} : g(z^*) \to z^* \quad \text{with}$$

$$g \circ \overline{g}_{z^*} = \text{inc}_{g(z^*)}, \tag{2.24}$$

the inclusion map of the subset $g(z^*)$ into $|K|$. Clearly, $\overline{g}_{z^*}$ is linear.

At the other extreme, g is called *totally degenerate* on z^*, or just z^* is totally degenerate, if $\dim z^* > 0$ but $\dim\, g(z^*) = 0$, i.e. the g maps the entire positive dimensional simplex z^* to a vertex.

In general, for $v \in V(g(z^*))$ the subset $(g|z^*)^{-1}(v)$ of z^* is either a vertex of z^* or a totally degenerate face of z^*. It is the face spanned by those vertices of z^* which map to v. Of course, for v_0, v_1 distinct vertices of $g(z^*)$, the faces $(g|z^*)^{-1}(v_0)$ and $(g|z^*)^{-1}(v_1)$ are disjoint. The set of vertices of z^* is thus partitioned by using the vertices of $g(z^*)$. We will exploit the fact that z^* is the polyhedral *join* of the associated faces.

First, define the *degeneracy cell* of z^* to be the product of these faces, with interior the product of the interiors:

$$T^{z^*} = \Pi\{(g|z^*)^{-1}(v) : v \in V(g(z^*))\}.$$

$$(T^{z^*})^\circ = \Pi\{((g|z^*)^{-1}(v))^\circ : v \in V(g(z^*))\}. \tag{2.25}$$

For $t \in T^{z^*}$ we write $t(v)$ for the element of $(g|z^*)^{-1}(v)$ associated with

$v \in V(g(z^*))$. Now we generalize (2.24) by defining for $y \in g(z^*)$ and $t \in T^{z^*}$

$$\overline{g}_{z^*}(y,t) = \sum_{v \in V(g(z^*))} c_K(y)_v t(v). \tag{2.26}$$

This yields a map

$$\overline{g}_{z^*} : g(z^*) \times T^{z^*} \to z^*, \quad \text{with}$$

$$g \circ \overline{g}_{z^*} = \text{proj}_{g(z^*)}, \tag{2.27}$$

the projection of the product to the first factor. Notice that z^* is nondegenerate exactly when every $(g|z^*)^{-1}(v)$ consists of a single vertex. In that case the degeneracy cell T^{z^*} is the single point associating to each vertex of $g(z^*)$ the vertex of z^* mapping to it. We identify $g(z^*)$ with its product with the singleton set T^{z^*}. This identifies the map $\overline{g}_{z^*}$ of (2.24) with that of (2.27) in the nondegenerate case.

2 Proposition. *The map $\overline{g}_{z^*} : g(z^*) \times T^{z^*} \to z^*$, with $x = \overline{g}_{z^*}(y,t)$, is C^∞, surjective and is linear in the y and t variables separately. For any x in z^*, $y = g(x)$ and so y is determined uniquely by x; for every $v \in V(g(z^*))$ such that $c_K(y)_v > 0$ the point $t(v)$ in $(g|z^*)^{-1}(v)$ is also uniquely determined by x. The restriction of $\overline{g}_{z^*}$ to $(g(z^*))^\circ \times (T^{z^*})^\circ$ is a C^∞ diffeomorphism onto $(z^*)^\circ$.*

Proof. Bilinearity and hence smoothness are clear. Now for each vertex v of $g(z^*)$ let I_v be the nonempty set of vertices of z^* mapping to v. Thus, I_v spans the face $(g|z^*)^{-1}(v)$ and $\{I_v : v \in V(g(z^*))\}$ is a partition of $V(z^*)$ For $x \in z^*$ and $y = g(x)$ we have from (2.12) and (2.21)

$$x = \sum_{v \in V(g(z^*))} \sum_{w \in I_v} c_{K^*}(x)_w w,$$

$$y = \sum_{v \in V(g(z^*))} c_K(y)_v v \quad \text{with}$$

$$c_K(y)_v = \sum_{w \in I_v} c_{K^*}(x)_w. \tag{2.28}$$

When $c_K(y)_v > 0$ we can define $t \in T^{z^*}$ by

$$t(v) = \frac{1}{c_K(y)_v} \sum_{w \in I_v} c_{K^*}(x)_w w, \tag{2.29}$$

and when $c_K(y)_v = 0$ we can choose $t(v)$ an arbitrary element of $(g|z^*)^{-1}(v)$. By (2.26) $\overline{g}_{z^*}(y,t) = x$ and so $\overline{g}_{z^*}$ is onto. Because $\{I_v\}$ forms a partition of the independent set $V(z^*)$, it is clear that $\overline{g}_{z^*}(y,t) = x$ implies $t(v)$ is given by (2.29) provided $c_K(y)_v > 0$. Clearly, from (2.27)

$$g(\overline{g}_{z^*}(y,t)) = y \tag{2.30}$$

and so y is determined by x as well.

Finally, it is clear that $x \in (z^*)^\circ$, i.e. $c_{K^*}(x)_w > 0$ for all $w \in V(z^*)$ iff $y \in (g(z^*))^\circ$ and $t(v) \in ((g|z^*)^{-1}(v))^\circ$ for all $v \in V(g(z^*))$. Smoothness of the inverse on $(z^*)^\circ$ follows from (2.29).

Remark. In particular, we observe that $c_{K^*}(x)_w > 0$ for all $w \in V(z^*)$ implies $c_K(y)_v > 0$ for all $v \in V(g(z^*))$, i.e.

$$x \in (z^*)^\circ \Rightarrow g(x) \in (g(z^*))^\circ. \tag{2.31}$$

□

Fixing $t \in T^{z^*}$, we define the linear map

$$\overline{g}_{z^*,t} : g(z^*) \to z^*$$

$$\overline{g}_{z^*,t}(y) = \overline{g}_{z^*}(y,t), \text{ so that}$$

$$g \circ \overline{g}_{z^*,t} = \mathrm{inc}_{g(z^*)}. \tag{2.32}$$

When z^* is degenerate the separate $\overline{g}_{z^*,t}$'s are not onto. We denote the image of $\overline{g}_{z^*,t}$ by $\langle z^*, t\rangle$. It is a simplex spanned by the independent set $\{t(v) : v \in V(g(z^*))\}$ and $\overline{g}_{z^*,t}$ regarded as a map from $g(z^*)$ to $\langle z^*, t\rangle$ is a linear isomorphism with inverse the restriction of g to $\langle z^*, t\rangle$.

If $z_1^* \in K^*$ and $t_1 \in T^{z_1^*}$ then we call (z^*, t) a *face* of (z_1^*, t_1), denoted $(z^*, t) < (z_1^*, t_1)$, if z^* is a face of z_1^* and the function t agrees with t_1 on the vertices of $g(z^*) \subset g(z_1^*)$, i.e.

$$(z^*, t) < (z_1^*, t_1) \Leftrightarrow$$

$$(2.33) \qquad z^* < z_1^* \text{ and } t(v) = t_1(v) \text{ for all } v \in V(g(z^*)).$$

3 Corollary. *Let $z^*, z_1^* \in K^*$ with $z^* < z_1^*$ and let $x \in |K^*|$.*

(a) For $t \in T^{z^*}$

$$(2.34) \qquad x \in \langle z^*, t\rangle \Leftrightarrow \overline{g}_{z^*,t}(g(x)) = x.$$

If $x \in (z^*)^\circ$ then there is a unique $t \in T^{z^*}$ such that $x \in \langle z^*, t\rangle$ and in that case $t \in (T^{z^*})^\circ$ and $x \in (\langle z^*, t\rangle)^\circ$.

(b) For $t \in T^{z^*}$ and $t_1 \in T^{z_1^*}$ the following conditions are equivalent:

(1) $(z^*, t) < (z_1^*, t_1)$.

(2) $\langle z^*, t\rangle$ is a face of the simplex $\langle z_1^*, t_1\rangle$.

(3) $\langle z^*, t\rangle$ is a subset of $\langle z_1^*, t_1\rangle$.

(4) $\langle z^*, t\rangle = \overline{g}_{z_1^*,t_1}(g(z^*))$.

(c) If $x \in \langle z^*, t\rangle$, $x \in \langle z_1^*, t_1\rangle$ and $x \in (z^*)^\circ$, then $(z^*, t) < (z_1^*, t_1)$.

Proof. (a) The equation in (2.34) implies x is in the image of $\overline{g}_{z^*,t}$. On the other hand, $x \in \langle z^*, t\rangle$ implies $x = \overline{g}_{z^*,t}(y)$ for some $y \in g(z^*)$ and by (2.32) $g(x) = y$. The rest follows directly from Proposition 2.

(b) (1) $\Rightarrow$ (2): By (2.33) the vertex set of $\langle z^*, t\rangle$, $\{t(v) : v \in V(g(z^*))\}$ is contained in $V(\langle z_1^*, t_1\rangle)$.

(2) $\Rightarrow$ (3): Obvious.

(3) $\Rightarrow$ (4): $g(\langle z^*, t\rangle) = g(z^*)$. Now apply (2.34) to this set of x's in the subset $\langle z^*, t\rangle$ of $\langle z_1^*, t_1\rangle$.

(4) $\Rightarrow$ (1): For each vertex v of $g(z_1^*)$, $t_1(v) = \overline{g}_{z_1^*,t_1}(v)$. But for $v \in V(g(z^*))$ $t(v) \in \langle z^*, t\rangle \subset \langle z_1^*, t_1\rangle$ and so by (2.34) $t(v) = \overline{g}_{z_1^*,t_1}(g(t(v))) = \overline{g}_{z_1^*,t_1}(v)$. That is, (2.33) holds.

(c) If $x \in (z^*)^\circ$ and $y = g(x)$ then $c_K(y)_v > 0$ for all $v \in V(g(z^*))$ and so $x \in \langle z^*, t\rangle$ and $x \in \langle z_1^*, t_1\rangle$ defines $t(v) = t_1(v)$ uniquely by (2.29) for all $v \in V(g(z^*))$. □

While $\langle z^*, t\rangle$ is a simplex contained in z^*, it is usually not an element of K^*. It is a simplex of K^* exactly when each $t(v)$ is a vertex of z^*. Define:

$$(2.35) \qquad V(T^{z^*}) = \Pi_{v\in V(g(z^*))} V((g|z^*)^{-1}(v)).$$

4 Corollary. *For $t \in T^{z^*}$, $\langle z^*, t\rangle$ is a simplex of K^*, and so is a face of z^*, iff $t \in V(T^{z^*})$. $\{\langle z^*, t\rangle : t \in V(T^{z^*})\}$ is the set of those faces if z^* on which g is nondegenerate with image all of $g(z^*)$. For $z_1^* = \langle z^*, t\rangle$ with $t \in V(T^{z^*})$ the linear isomorphism $\overline{g}_{z^*,t} : g(z^*) \to \langle z^*, t\rangle$, defined by (2.32), and the linear isomorphism $\overline{g}_{z_1^*} : g(z_1^*) \to z_1^*$, defined by (2.24), are the same. Finally, for any $t \in T^{z^*}$, linear map $\overline{g}_{z^*,t} : g(z^*) \to z^*$ is a convex combination of the maps $\overline{g}_{z_1^*} : g(z^*) \to z^*$ as z_1^* varies over the set of faces $\{\langle z^*, t\rangle : t \in V(T^{z^*})\}$.*

Proof. The first part is obvious from the above definitions and the last result follows from linearity of $\overline{g}(y, t)$ in t. □

The simplicial complex K^* is called a *subdivision* of the complex K if the two triangulate the same polyhedron and, regarded as covers of this polyhedron, K^* refines K. That is, K^* and K satisfy:

$$|K^*| = |K|. \tag{2.36}$$

$$\text{For every } z^* \in K^*, \text{ there exists } z \in K \text{ such that } z^* \subset z. \tag{2.37}$$

Thus, each $z^* \in K^*$ has a carrier in K, the smallest simplex in K containing z^*, denoted $j(z^*)$.

5 Lemma. *Assume K^* is a subdivision of K, $z^* \in K^*$ and $x \in (z^*)^\circ$. The following conditions on $z \in K$ are equivalent:*

(1) z is the carrier of z^*, i.e. $z = j(z^*)$.

(2) z is the carrier in K of $V(z^*)$.

(3) $(z^*)^\circ \subset (z)^\circ$.

(4) z is the carrier of x, i.e. $z = q(x)$.

Proof. (1) ⇒ (2): By convexity, z contains z^* iff z contains $V(z^*)$.

(2) ⇒ (3): If z is the carrier of the finite set $V(z^*)$ then for every $v \in V(z)$ there exists $w \in V(z^*)$ such that $c_K(w)_v > 0$. Hence, if $y \in (z^*)^\circ$, $c_K(y)_v > 0$ for all $v \in V(z)$ and so $y \in (z)^\circ$.

(3) ⇒ (4): $x \in (z^*)^\circ$ implies $x \in (z)^\circ$ by (3) and so $q(x) = z$.

(4) ⇒ (1): We have just proved that $j(z^*)$ is the carrier of x, i.e. $j(z^*) = q(x)$. Hence, (1) follows from (4). □

By analogy with (2.17) we define the carrier relations j, $J \subset K^* \times K$:

$$j = \{(z^*, z) : (z^*)^\circ \subset (z)^\circ\}$$

$$J = \{(z^*, z) : z^* \subset z\}. \tag{2.38}$$

We leave to the reader the proof of the following easy consequences of the definitions.

6 Lemma. *(a) For $(z^*, z) \in K^* \times K$, $z \in J(z^*)$ iff $j(z^*)$ is a face of z.*
(b) If $z \in J(z^*)$ then

$$\dim z \geq \dim j(z^*) \geq \dim z^* \tag{2.39}$$

and $\dim z = \dim j(z^*)$ *iff* $z = j(z^*)$. *In particular, dim* $z =$ *dim* z^* *implies* $z = j(z^*)$.
(c) If z_2^* is a face of z_1^* in K^* then $j(z_2^*)$ is a face of $j(z_1^*)$ in K. □

If L is a subcomplex of K then $J^{-1}(L)$ is the subcomplex of K^* which subdivides L, i.e.

$$|J^{-1}(L)| = |L|. \tag{2.40}$$

Since the subdivision of a k dimensional complex is k dimensional we have:

$$J^{-1}(\overline{S}^k(K)) \subset \overline{S}^k(K^*)$$

$$|\overline{S}^k(K)| \subset |\overline{S}^k(K^*)|. \tag{2.41}$$

In particular, for $k = 0$ we have

$$V(K) \subset V(K^*). \tag{2.42}$$

7 Definition. *Let K be a simplicial complex and A be a closed subset of the polyhedron $|K|$. A is said to be* properly included *in K if A does not meet any pair of disjoint simplices of K. That is, for all $z_0, z_1 \in K$*

$$A \cap z_0 \neq \emptyset \text{ and } A \cap z_1 \neq \emptyset \Rightarrow z_0 \cap z_1 \neq \emptyset. \tag{2.43}$$

A subdivision K^* of K is called a proper subdivision *of K (or just: K^* is proper for K) if every simplex of K^* is properly included in K.* □

8 Lemma. *Let K be a simplicial complex, K^* be a subdivision of K and A, $\tilde{A}$ be closed subsets of $|K|$.*

(a) If A is a singleton set then A is properly included in K.

(b) If A is properly included in K and $\tilde{A} \subset A$ then $\tilde{A}$ is properly included in K.

(c) If A is properly included in K^* then A is properly included in K.

(d) If K^{**} is a subdivision of K^* and either K^{**} is proper in K^* or K^* is proper in K then K^{**} is proper in K.

(e) If A is contained in the open star $St(z)$ of some simplex z of K then A is properly included in K.

(f) Let z^* be a simplex of K^* with dim $z^* = 1$. The subset z^* is properly included in K iff z^* is contained in the open star $St(v)$ of some vertex v of K.

(g) For a simplex z^* of K^* the following conditions are equivalent:

(1) z^* is properly included in K.

(2) If z_0^* and z_1^* are faces of z^* then the carriers $j(z_0^*)$ and $j(z_1^*)$ have a common face in K.

(3) Every face z_0^* of z^* with dim $z_0^* = 1$ is properly included in K.

(h) Any derived subdivision of K is proper for K.

Proof. (a) and (b) are obvious.

(c) Let $x_\alpha \in A \cap z_\alpha$ $(\alpha = 0, 1)$ with $z_0, z_1 \in K$. If z_α^* is the K^* carrier of x_α then $z_0^* \cap z_1^* \neq \emptyset$ because A is properly included in K^*. Since $z_\alpha^* \subset z_\alpha$ $(\alpha = 0, 1)$ it follows that $z_0 \cap z_1 \neq \emptyset$. (d) clearly follows from (b) and (c).

(e) If $z_0 \cap A \neq \emptyset$, $z_1 \cap A \neq \emptyset$ and $A \subset \mathrm{St}(z)$ then $z < z_0$ and $z < z_1$ by (2.16). So $z \subset z_0 \cap z_1$.

Now suppose that $z^* \in K^*$ is properly included in K and z_0^*, $z_1^* < z^*$. The carriers $z_\alpha = j(z_\alpha^*)$ $(\alpha = 0, 1)$ meet z^* and so intersect in a common face. This proves (1) $\Rightarrow$ (2) of (g). Now suppose that z_α^* is a vertex v_α^* of K^* $(\alpha = 0, 1)$ and that z^* is spanned by v_0^* and v_1^*. Let v be a vertex of $z_0 \cap z_1$. For $\alpha = 0, 1$ $v_\alpha^* \in (z_\alpha)^\circ \subset \mathrm{St}(v)$. The remaining points of z^* are in $(z^*)^\circ$ which is contained in $(z)^\circ$ where $z = j(z^*)$. Since $z_\alpha < z$ $(\alpha = 0, 1)$ it follows that $(z^*)^\circ \subset (z)^\circ \subset \mathrm{St}(v)$. Together with (e) this proves (f). This also proves (2) $\Rightarrow$ (3) of (g).

(g) (3) $\Rightarrow$ (1): If $z_\alpha \cap z^* \neq \emptyset$ then the intersection is triangulated by a subcomplex of K^* and so there is some vertex v_α^* of K^* contained in $z_\alpha \cap z^*$

($\alpha = 0, 1$). The span $[v_0^*, v_1^*]$ is a face of z^* of dimension at most 1 so by assumption (3) (or (a)) it is properly contained in K. Hence $z_0 \cap z_1 \neq \emptyset$.

(h) A typical simplex of a first derived subdivision of K is of the form $[b(z_0), \dots, b(z_k)]$ with $z_0 < \dots < z_k$ a strictly increasing sequence of faces in K and with $b(z_i) \in (z_i)^\circ$ for $i = 0, 1, \dots, k$ (see Appendix: Chapter 10). So the simplex is contained in $\mathrm{St}(z_0)$. It follows from (e) that the first derived is a proper subdivision of K. Any further subdivision is proper by (d). □

It follows that the condition that K^* be a proper subdivision is a rather mild restriction.

The major result of this section comes from an estimate for stochastic matrices.

An $(m+1) \times (n+1)$ matrix P is called *stochastic* when it is nonnegative and each column sum is 1. That is, each column vector lies in the unit simplex Δ_{m+1} of $\mathbf{R}^{m+1}$ (the notation follows (2.1) and (2.2) using the natural number $m+1$ to stand for the set $\{0, 1, \dots, m\}$). Thus, P satisfies

$$\sum_i P_{ij} = 1 \text{ for all } j \text{ and } 1 \geq P_{ij} \geq 0 \text{ for all } i, j. \tag{2.44}$$

As usual P induces a linear map from $\mathbf{R}^{n+1}$ to $\mathbf{R}^{m+1}$ by $a \mapsto Pa$. When P is stochastic, it maps Δ_{n+1} to Δ_{m+1} and $\mathbf{R}_0^{n+1}$ to $\mathbf{R}_0^{m+1}$. For a stochastic matrix, P, we let $\| P \|$ denote the norm of P regarded as a linear map from $\mathbf{R}_0^{n+1}$ to $\mathbf{R}_0^{m+1}$. As usual, we use the l^1 norm on the coordinate spaces. Thus,

$$\| P \| = \sup\{\| Pa \| : a \in \mathbf{R}_0^{n+1} \text{ and } \| a \| \leq 1\}. \tag{2.45}$$

9 Proposition. *If P is an $(m+1) \times (n+1)$ stochastic map, then $\| P \| \leq 1$ and if $n = 0$ then $\| P \| = 0$.*

Let $\theta > 0$ and assume that $n > 0$. If for every pair $j_1, j_2 \in \{0, \dots n\}$ there exists $i_0 \in \{0, \dots m\}$ such that

$$P_{i_0 j_1} \geq \theta \text{ and } P_{i_0 j_2} \geq \theta, \tag{2.46}$$

then

$$\| P \| \leq 1 - (\theta/n). \tag{2.47}$$

Proof. If $n = 0$ then $\mathbf{R}_0^{n+1} = \{0\}$ and so $P = 0$ on $\mathbf{R}_0^{n+1}$. In general,

$$\| Pa \| = \sum_i |\sum_j P_{ij} a_j|$$

$$\leq \sum_{i,j} P_{ij}|a_j| = \sum_j |a_j| = \| a \| . \tag{2.48}$$

Now assume $a \in \mathbf{R}_0^{n+1}$ and $a \neq 0$. Let a_+ be the sum of the positive components of a and a_- the sum of the negative components. Clearly,

$$a_+ - a_- = \| a \| \quad \text{and} \quad a_+ + a_- = 0. \tag{2.49}$$

Hence, $a_+ = -a_- = \|a\| /2$. Furthermore, there are at most n components of either type. It follows that there exist $j_1, j_2 \in \{0, \ldots, n\}$ such that

$$a_{j_1} \geq \| a \| /2n; \quad -a_{j_2} \geq \| a \| /2n. \tag{2.50}$$

If (2.46) holds then

$$2P_{i_0 j_1} a_{j_1} \geq \theta \| a \| /n;$$

$$-2P_{i_0 j_2} a_{j_2} \geq \theta \| a \| /n. \tag{2.51}$$

From which we obtain

$$P_{i_0 j_1}|a_{j_1}| + P_{i_0 j_2}|a_{j_2}| - \theta \| a \| /n =$$

$$P_{i_0 j_1} a_{j_1} - P_{i_0 j_2} a_{j_2} - \theta \| a \| /n \geq$$

$$\max(P_{i_0 j_1} a_{j_1} + P_{i_0 j_2} a_{j_2}, -P_{i_0 j_1} a_{j_1} - P_{i_0 j_2} a_{j_2})$$

$$= |P_{i_0 j_1} a_{j_1} + P_{i_0 j_2} a_{j_2}|. \tag{2.52}$$

So we can adjust (2.48) in this case to get

$$\| Pa \| = \sum_i |\sum_j P_{ij} a_j| \leq$$

$$\sum_{ij} P_{ij}|a_j| - \theta \| a \| /n = (1 - \theta/n) \| a \| . \tag{2.53}$$

□

The dimension of K, i.e. the $\max\{\dim z : z \in K\}$, is the same as the dimension of any subdivision of K. We write the dimension as dim $|K|$. A more technical constant will also be needed.

10 Definition. *Let K be a simplicial complex and K^* be a subdivision of K. The* separation constant *of K^* in K, denoted $\theta(K^*, K)$, is defined to be*

$$\min\{c_K(w)_v : w \in V(K^*), v \in V(K) \text{ and } c_K(w)_v > 0\}. \tag{2.54}$$

That is, $\theta(K^*, K)$ is the minimum positive barycentric coordinate in K of a vertex of K^*. □

11 Theorem. *Let K be a simplicial complex of dimension $d > 0$, let K^* be a subdivision of K with separation constant θ, and let $g : K^* \to K$ be a simplicial map.*

For $z^* \in K^*$ and $t \in T^{z^*}$ the linear map $\overline{g}_{z^*,t}$ is distance nonincreasing with respect to the metric d_K on $|K|$. That is,

$$d_K(\overline{g}_{z^*,t}(y_1), \overline{g}_{z^*,t}(y_2)) \le d_K(y_1, y_2) \tag{2.55}$$

for all $y_1, y_2 \in g(z^*)$. If z^* is properly included in K then $\overline{g}_{z^*,t}$ is a contraction. In fact,

$$d_K(\overline{g}_{z^*,t}(y_1), \overline{g}_{z^*,t}(y_2)) \le (1 - \frac{\theta}{d}) d_K(y_1, y_2). \tag{2.56}$$

Proof. List the vertices of K beginning with those of $g(z^*)$ so the $V(K) = \{v_0, \ldots, v_m\}$ and $V(g(z^*)) = \{v_0, \ldots, v_n\}$. Define the $(m+1) \times (n+1)$ stochastic matrix P:

$$P_{ij} = c_K(\overline{g}_{z^*,t}(v_j))_{v_i}. \tag{2.57}$$

That is, column j of P is the K barycentric coordinate vector of $\overline{g}_{z^*,t}(v_j)$. For $y_1, y_2 \in g(z^*)$ let a, b be the respective $g(z^*)$ coordinate vectors.

$$a = c_{g(z^*)}(y_1) \quad \text{and} \quad b = c_{g(z^*)}(y_2). \tag{2.58}$$

By (2.15)

$$d_K(y_1, y_2) = \| a - b \| . \tag{2.59}$$

By linearity of $\overline{g}_{z^*,t}$:

$$Pa = c_K(\overline{g}_{z^*,t}(y_1)) \text{ and } Pb = c_K(\overline{g}_{z^*,t}(y_2)). \tag{2.60}$$

So by (2.15) again and (2.3):

$$d_K(\overline{g}_{z^*,t}(y_1), \overline{g}_{z^*,t}(y_2)) = \| Pa - Pb \| \leq \| P \| \, \| a - b \| . \tag{2.61}$$

Thus, (2.55) follows from the first paragraph of Proposition 9.

Now assume that z^* is included properly in K. We observe first that by linearity of $\overline{g}_{z^*,t}$ in t, i.e. by Corollary 4, we can reduce to the case where t is a vertex of T^{z^*}. When $t \in V(T^{z^*})$ and $j_1, j_2 \in \{0, \ldots, n\}$ then $\overline{g}_{z^*,t}(v_{j_\alpha}) = t(v_{j_\alpha})$ is a vertex of z^* for $\alpha = 1, 2$. By Lemma 8f there exists $i_0 \in \{0, \ldots, m\}$ such that $t(v_{j_1})$ and $t(v_{j_2})$ are both contained in the star of the vertex v_{i_0}. That is, the v_{i_0} barycentric coordinate of $t(v_{j_\alpha})$ is positive for $\alpha = 1, 2$. By definition of the separation constant $\theta(K^*, K)$ this says that

$$P_{i_0 j_\alpha} = c_K(t(v_{j_\alpha}))_{v_{i_0}} \geq \theta \;\; (\alpha = 1, 2). \tag{2.62}$$

Since $d \geq n$, (2.56) now follows from (2.59), (2.61) and (2.47).

Remark. In the trivial case $d = 0$, $K^* = K$ and each $\overline{g}_{z^*}$ is a single point map. □

3. The Shift Factor Maps for a Simplicial Dynamical System

Throughout this chapter X is a polyhedron triangulated by a simplicial complex K, K^* is a subdivision of K and $g : K^* \to K$ is a simplicial map. Thus, $X = |K| = |K^*|$ and $g : X \to X$ is the p.l. map obtained from the simplicial map. We call $g : K^* \to K$ a *simplicial dynamical system* if, in addition, K^* is a proper subdivision of K, i.e. no simplex of K^* meets a disjoint pair of simplices of K (see Definition 2.7). In addition to the simplicial map g we have the carrier map $j : K^* \to K$ and the inclusion relation $J : K^* \to K$ (see (2.38)).

We want to analyze the dynamics of the continuous map g on X by using the finite set map $g : K^* \to K$. Of course, here we can't iterate directly, but for $z_1^* \in K^*$, $g(z_1^*) = z_2$ is triangulated by the subcomplex $J^{-1}(z_2)$ in K^*. We can choose $z_2^* \in J^{-1}(z_2)$ and apply g again. Similarly, given $z_1 \in K$, $J^{-1}(z_1)$ is a subcomplex of K^* and so $g(J^{-1}(z_1))$ is a subcomplex of K. Choose $z_2 \in g(J^{-1}(z_1))$ and apply g again. Thus, we are in the two-alphabet situation described in Proposition 1.4 with the map $g : K^* \to K$ and the relation $J : K^* \to K$. Define, as in (1.20),

$$G^* = J^{-1} \circ g = \{(z_1^*, z_2^*) : z_2^* \subset g(z_1^*)\} \subset K^* \times K^*$$

$$G = g \circ J^{-1} = \{(z_1, z_2) : z_2 \subset g(z_1)\} \subset K \times K. \tag{3.1}$$

Over X lies the sample path space $X_g \subset X^{\mathbf{Z}}$ on which the shift homeomorphism s_g acts. The projection $\pi_0 : X_g \to X$ associating to the sequence ξ in X_g the point ξ_0 in X, maps s_g to g. Recall that, because g is a map, the analogue π_0 on the one-sided sample path space X_g^+ identifies s_g^+ with g.

We will relate g on X and s_g on X_g with the shift maps $s_{G^*}^+$ on $K^{*+}_{G^*}$ and s_{G^*} on $K^*_{G^*}$. For any subset L of K^* we can restrict G^* to L, i.e. consider the relation $G^* \cap (L \times L)$. The associated sample path spaces are the subspaces given by

$$L_{G^*} = K^*_{G^*} \cap (L^{\mathbf{Z}})$$

$$L^+_{G^*} = K^{*+}_{G^*} \cap (L^{\mathbf{Z}_+}). \tag{3.2}$$

Recall that for each $z^* \in K^*$ we defined the degeneracy cell T^{z^*}, a product of certain faces of z^*, see (2.25). The cell T^{z^*} consists of a single point exactly when g is nondegenerate on z^*. To deal with degenerate simplices we define

$$\hat{K} = \cup\{\{z^*\} \times T^{z^*} : z^* \in K^*\} \tag{3.3}$$

Thus, $\hat{K}$ expands the finite set K^* to a disjoint union of closed cells, blowing up the point z^* to its degeneracy cell. We let $\hat{p}$ denote the projection to the first coordinate

$$\hat{p} : \hat{K} \to K^*, \text{ with } \hat{p}(z^*, t) = z^*. \tag{3.4}$$

We extend the closed relation G^* on K^* to the closed relation $\hat{G}$ on $\hat{K}$ by preimage:

$$\begin{aligned} &\hat{G} = (\hat{p} \times \hat{p})^{-1}(G^*) \subset \hat{K} \times \hat{K} \\ &(z_1^*, t_1) \in \hat{G}(z_0^*, t_0) \Leftrightarrow z_1^* \in G^*(z_0^*). \end{aligned} \tag{3.5}$$

If we use the continuous map $\hat{p}$ on each coordinate we obtain:

$$\begin{aligned} &\hat{p} : \hat{K}^{\mathbf{Z}} \to K^{*\mathbf{Z}} \\ &\hat{p}^+ : \hat{K}^{\mathbf{Z}_+} \to K^{*\mathbf{Z}_+}, \end{aligned} \tag{3.6}$$

then the $\hat{G}$ sample path spaces are just the preimages of G^* sample path spaces

$$\begin{aligned} &\hat{K}_{\hat{G}} = \hat{p}^{-1}(K^*_{G^*}) \\ &\hat{K}^+_{\hat{G}} = \hat{p}^{-1}(K^{*+}_{G^*}). \end{aligned} \tag{3.7}$$

We will use the notation $\mathbf{z}^*$ for sequences in $L^{*\mathbf{Z}}$ and $K^{*\mathbf{Z}_+}$ with $\mathbf{z}$ and $(\mathbf{z}^*, \mathbf{t})$ for sequences in $K^{\mathbf{Z}}$, $K^{\mathbf{Z}_+}$ and $\hat{K}^{\mathbf{Z}}$. $\hat{K}^{\mathbf{Z}_+}$. We will drop the boldface type for the coordinates: z_i^*, z_i, (z_i^*, t_i), $i \in \mathbf{Z}$, $\mathbf{Z}_+$ of the corresponding sequences. Recall that for each $(z^*, t) \in \hat{K}$ we have defined the simplex $\langle z^*, t\rangle \subset z^*$. It is the image of the linear map $\overline{g}_{z^*,t} : g(z^*) \to z^*$, or, equivalently, it is the simplex spanned by $\{t(v) : v \in V(g(z^*))\}$.

Recall the carrier maps $q : X \to K$ and $q^* : X \to K^*$ defined by (2.17) and related by Lemma 2.5. For $(x, z^*) \in X \times K^*$ we have

$$q^*(x) = z^* \Leftrightarrow x \in (z^*)^\circ. \tag{3.8}$$

By Corollary 2.3, there is for $z^* = q^*(x)$ a unique $t \in T^{Z^*}$ such that $x \in \langle z^*, t\rangle$ or, equivalently,

$$x = \overline{g}_{z^*,t}(g(x)). \tag{3.9}$$

Thus, we define the carrier map

$$\hat{q} : X \to \hat{K}$$

(3.10) $$\hat{q}(x) = (z^*, t) \Leftrightarrow x \in \langle z^*, t\rangle \cap (z^*)^\circ.$$

Clearly

(3.11) $$\hat{p} \circ \hat{q} = q^* : X \to K^*.$$

We define the incidence relation on $\hat{K}$ by (2.33), $(z^*, t) < (z_1^*, t_1)$ when $z^* < z_1^*$ and $t(v) = t_1(v)$ for all $v \in V(g(z^*))$. Then we extend (2.18)

$$x \in \langle z_1^*, t_1\rangle \Leftrightarrow \hat{q}(x) < (z_1^*, t_1), \text{ in which case}$$

(3.12) $$\dim q(x) = \dim z_1^* \text{ iff } \hat{q}(x) = (z_1^*, t_1).$$

The first equivalence follows from Corollary 2.3c. Then by (2.18), $\dim q(x) = \dim z_1^*$ iff $q(x) = z_1^*$ in which case by Corollary 2.3c again (z_1^*, t_1) is a face of, and so equals, $\hat{q}(x)$.

1 Lemma. *Assume $x \in (z^*)^\circ$. If $z = g(z^*)$ then $g(x) \in (z)^\circ$. That is, the following diagram commutes*

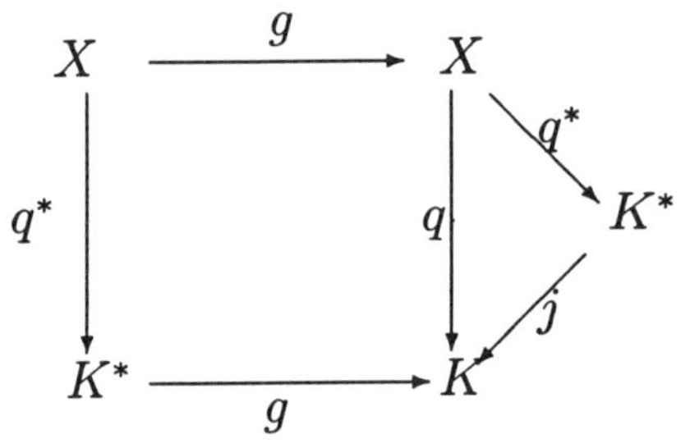

(3.13)

Proof. Even if g is a degenerate on z^* it is still true that g maps interior points of z^* to interior points of $g(z^*) = z$, see (2.31). Thus, the left hand rectangle commutes. The triangle commutes by Lemma 2.5. □

By Lemma 1 any g chain is mapped by q^* to a G^* chain and so by $\hat{q}$ to a $\hat{G}$ chain. Thus, we can define functions:

$$q^* : X_g \to K^*_{G^*}\ ;\ \hat{q} : X_g \to \hat{K}_{\hat{G}}$$
$$q^*(\xi)_i = q^*(\xi_i); \text{and } \hat{q}(\xi)_i = \hat{q}(\xi_i) \text{ for; all } i \in \mathbf{Z}$$
$$q^{*+} : X \to K^{*+}_{G^*}\ ;\ \hat{q}^+ : X \to \hat{K}^+_{\hat{G}}$$
$$q^{*+}(x)_i = q^*(g^i(x)) \text{ and } \hat{q}(x)_i = \hat{q}(g^i(x)) \text{ for all } i \in \mathbf{Z}_+. \tag{3.14}$$

Now we define the relations

$$\hat{h} = \{((\mathbf{z}^*, \mathbf{t}), \xi) \in \hat{K}_{\hat{G}} \times X_g : \xi_i \in \langle z^*_i, t_i\rangle \text{ for all } i \in \mathbf{Z}\}.$$
$$\hat{h}^+ = \{((\mathbf{z}^*, \mathbf{t}), x) \in \hat{K}^+_{\hat{G}} \times X : g^i(x) \in \langle z^*_i, t_i\rangle \text{ for all } i \in \mathbf{Z}_+\}. \tag{3.15}$$

2 Theorem. *Let $g : K^* \to K$ be a simplicial map with K^* a subdivision of K and let $g : X \to X$ be the associated p.l. map where X is the polyhedron $|K| = |K^*|$. Define the sample path relations $\hat{p}$, $\hat{p}^+$, $\hat{q}$, $\hat{q}^+$, $\hat{h}$ and $\hat{h}^+$ as above. Let s_{G^*}, $s_{\hat{G}}$ and s_g denote the sample path shift homeomorphisms and $s^+_{G^*}$, $s^+_{\hat{G}}$ be the one-sided sample path shift maps. Let $\pi^+_{G^*} : K^*_{G^*} \to K^{*+}_{G^*}$, $\pi^+_{\hat{G}} : \hat{K}_{\hat{G}} \to \hat{K}^+_{\hat{G}}$ and $\pi_0 : X_g \to X$ be the projection maps.*

(a) The mappings $\hat{p} : \hat{K}_{\hat{G}} \to K^*_{G*}$ and $\hat{p}^+ : \hat{K}^+_{\hat{G}} \to K^{*+}_{G^*}$ are continuous, open surjections satisfying:

$$s_{G^*} \circ \hat{p} = \hat{p} \circ s_{\hat{G}},$$
$$s^+_{G^*} \circ \hat{p}^+ = \hat{p}^+ \circ s^+_{\hat{G}},$$
$$\pi^+_{G^*} \circ \hat{p} = \hat{p}^+ \circ \pi^+_{\hat{G}}. \tag{3.16}$$

(b) The mappings $\hat{q} : X_g \to \hat{K}_{\hat{G}}$ and $\hat{q}^+ : X \to \hat{K}^+_{\hat{G}}$ are usually discontinuous but satisfy:

$$s_{\hat{G}} \circ \hat{q} = \hat{q} \circ s_g$$
$$s^+_{\hat{G}} \circ \hat{q}^+ = \hat{q}^+ \circ g$$
$$\pi^+_{\hat{G}} \circ \hat{q} = \hat{q}^+ \circ \pi_0. \tag{3.17}$$

(c) The relations $\hat{h} : \hat{K}_{\hat{G}} \to X_g$ and $\hat{h}^+ : \hat{K}^+_{\hat{G}} \to X$ are closed, surjective relations satisfying

$$s_g \circ \hat{h} = \hat{h} \circ s_{\hat{G}}$$

$$g \circ \hat{h}^+ = \hat{h}^+ \circ s^+_{\hat{G}}$$

(3.18) $$\pi_0 \circ \hat{h} = \hat{h}^+ \circ \pi^+_{\hat{G}}.$$

Furthermore,

(3.19) $$\hat{q} \subset \hat{h}^{-1} \text{ and } \hat{q}^+ \subset (\hat{h}^+)^{-1}.$$

(d) If $g : K^* \to K$ is a simplicial dynamical system, i.e. K^* is a proper subdivision of K then $\hat{h}$ and $\hat{h}^+$ are continuous, surjective mappings satisfying:

(3.20) $$\hat{h} \circ \hat{q} = 1_{X_g} \text{ and } \hat{h}^+ \circ \hat{q}^+ = 1_X.$$

Proof. (a) As the product of continuous, open surjections $\hat{p} : \hat{K}^{\mathbf{Z}} \to K^{+\mathbf{Z}}$ is a continuous, open surjection. For any subset A of $K^{*\mathbf{Z}}$ the restriction $\hat{p} : (\hat{p})^{-1}(A) \to A$ is hence a continuous, open surjection. The results for $\mathbf{Z}_+$ are similar and the identities of (3.16) are clear.

(b) The identities of (3.17) are obvious from the definitions (3.13) and (3.14).

(c) By (3.11), $((\mathbf{z}^*, \mathbf{t}), \xi) \in h$ iff

(3.21) $$\overline{g}_{z_i^*}(\xi_{i+1}, t_i) = \xi_i \text{ for all } i \in \mathbf{Z}$$

and similarly $((\mathbf{z}^*, \mathbf{t}), x) \in h^+$ iff

(3.22) $$\overline{g}_{z_i^*}(g^{i+1}(x), t_i) = g^i(x) \text{ for all } i \in \mathbf{Z}_+.$$

By continuity of the $\overline{g}_{z_i^*}$ maps, i.e. Proposition 2.2, these are closed conditions and so $\hat{h}$ and $\hat{h}^+$ are closed relations. The inclusions (3.19), e.g. $(\hat{q}(\xi), \xi) \in \hat{h}$, follow by reflecting upon the definitions (3.12) - (3.15). In particular, we have $\hat{h}(\hat{K}_{\hat{G}}) = X_g$ and $\hat{h}^+(\hat{K}^+_{\hat{G}}) = X$.

The first two composition equations of (3.18) follow from:

$$\xi \in \hat{h}(\mathbf{z}^*, \mathbf{t}) \Leftrightarrow s_g(\xi) \in \hat{h}(s_{\hat{G}}(\mathbf{z}^*, \mathbf{t}))$$

(3.23) $$x \in \hat{h}^+(\mathbf{z}^*, \mathbf{t}) \Leftrightarrow g(x) \in \hat{h}(s^+_{\hat{G}}(\mathbf{z}^*, \mathbf{t})) \text{ and } x = \overline{g}_{z_0^*, t_0}(g(x)).$$

Furthermore, clearly

$$\xi \in \hat{h}(\mathbf{z}^*, \mathbf{t}) \Rightarrow \xi_0 \in \hat{h}^+(\pi^+_{\hat{G}}(\mathbf{z}^*, \mathbf{t})) \tag{3.24}$$

proving $\pi_0 \circ \hat{h} \subset \hat{h}^+ \circ \pi^+_{\hat{G}}$. On the other hand, if $(\mathbf{z}^*, \mathbf{t}) \in \hat{K}_{\hat{G}}$ and $x \in \hat{h}^+(\pi^+_{\hat{G}}(\mathbf{z}^*, \mathbf{t}))$ then by defining

$$\xi_i = g^i(x) \text{ for } i \geq 0$$

$$\xi_i = \overline{g}_{z_i^*, t_i}(\xi_{i+1}) \text{ inductively for } i < 0 \tag{3.25}$$

we obtain the unique $\xi \in X_g$ such that

$$\xi_0 = x \text{ and } \xi \in \hat{h}(\mathbf{z}^*, \mathbf{t}). \tag{3.26}$$

This proves the third equation of (3.18).

Now for $(\mathbf{z}^*, \mathbf{t}) \in \hat{K}^+_{\hat{G}}$ define $z_i = g(z^*_{i-1})$ for $i = 1, 2, \ldots$. Composing each $\overline{g}_{z^*_{i-1}, t_{i-1}} : z_i \to z^*_{i-1}$ with the inclusion of z^*_{i-1} into z_{i-1} we can define the linear map for $k = 1, 2, \ldots$

$$\overline{g}^k_{(\mathbf{z}^*, \mathbf{t})} : z_k \to z^*_0$$

$$\overline{g}^k_{(\mathbf{z}^*, \mathbf{t})} = \overline{g}_{z^*_0, t_0} \circ \ldots \circ \overline{g}_{z^*_{k-1}, t_{k-1}}. \tag{3.27}$$

The sequence of images

$$\{\overline{g}^k_{(\mathbf{z}^*, \mathbf{t})}(z_k) : k = 1, 2, \ldots\} \tag{3.28}$$

is a decreasing sequence of nonempty, compact convex sets in z^*_0. Let x be a point in the intersection. By (2.32) $g^i(x)$ is in the image of $\overline{g}_{z^*_i, t_i}$ for $i = 0, 1, \ldots$ and so $x \in \hat{h}^+(\mathbf{z}^*, \mathbf{t})$. This completes the proof that $\hat{h}^+$ is a surjective relation. Now if $(\mathbf{z}^*, \mathbf{t}) \in \hat{K}_{\hat{G}}$ we obtain $x \in \hat{h}^+(\pi^+_{\hat{G}}(\mathbf{z}^*, \mathbf{t}))$ and then construct ξ following (3.25) to satisfy (3.26). Thus, $\hat{h}$ is a surjective relation.

(d) If K^* is a proper subdivision of K then by Theorem 2.11 the d_K diameter of the image $\overline{g}^k_{(\mathbf{z}^*, \mathbf{t})}(z_k)$ is at most $(1 - (\theta/d))^k$ times 2. Thus, for $(\mathbf{z}^*, \mathbf{t}) \in \hat{K}^+_{\hat{G}}$ there is a unique point $x \in X$ such that $x \in \hat{h}^+(\mathbf{z}^*, \mathbf{t})$. That is, $\hat{h}^+$ is a mapping. By the construction of (3.25) we see that $\hat{h}$ is a mapping. Because $\hat{h}^+$ and $\hat{h}$ are closed surjective relations they are continuous, surjective maps. From (3.19) we obtain $\hat{h} \circ \hat{q} \subset \hat{h} \circ \hat{h}^{-1}$ and this is contained in 1_{X_g} because $\hat{h}$ is a map. A relation inclusion between maps is an equality. The first equation in (3.20) follows and the second is obtained similarly. □

In order to prove part (d) we only needed that the diameters of the cells in the sequence (3.28) tended to zero. The full strength of Theorem 2.11 was not needed, but it is required for the following *Partial Shadowing Lemma.*

3 Theorem. *Let $g : K^* \to K$ be a simplicial dynamical system with $X = |K| = |K^*|$. Let θ be the separation constant of K^* in K and d be the dimension of X.*

(a) Assume $(\mathbf{z}^*, \mathbf{t}) \in \hat{K}^+_{\hat{G}}$ and $\eta \in X^{\mathbf{Z}_+}$ such that

$$\eta_i \in \langle z_i^*, t_i \rangle \quad \text{for all } i \in \mathbf{Z}_+ \tag{3.29}$$

If for some $\epsilon \geq 0$, η is an ϵ chain for g, i.e.

$$d_K(g(\eta_i), \eta_{i+1}) \leq \epsilon \quad \text{for all } i \in \mathbf{Z}_+ \tag{3.30}$$

then with $x = \hat{h}^+(\mathbf{z}^*, \mathbf{t})$,

$$d_K(\eta_i, g^i(x)) \leq (d/\theta)\epsilon \quad \text{for all } i \in \mathbf{Z}_+. \tag{3.31}$$

(b) Assume $(\mathbf{z}^*, \mathbf{t}) \in \hat{K}_{\hat{G}}$ and $\eta \in X^{\mathbf{Z}}$ such that

$$\eta_i \in \langle z_i, t_i \rangle \quad \text{for all } i \in \mathbf{Z}. \tag{3.32}$$

If for some $\epsilon \geq 0$, η is an ϵ chain for g, i.e.

$$d_K(g(\eta_i), \eta_{i+1}) \leq \epsilon \quad \text{for all } i \in \mathbf{Z} \tag{3.33}$$

then with $\xi = \hat{h}(\mathbf{z}^*, \mathbf{t})$,

$$d_K(\eta_i, \xi_i) \leq (d/\theta)\epsilon \quad \text{for all } i \in \mathbf{Z}. \tag{3.34}$$

Proof: (a) Using the notation of (3.27) we see that

$$x^k \equiv \overline{g}^k_{(\mathbf{z}^*, \mathbf{t})}(\eta_k) \in \overline{g}^k_{(\mathbf{z}^*, \mathbf{t})}(z_k) \tag{3.35}$$

for $k = 1, 2, \ldots$ and so, as x is the unique intersection point,

$$\text{Lim}_{k\to\infty}\{x^k\} = x. \tag{3.36}$$

Since $\eta_i \in \langle z_i^*, t_i\rangle$ we have by (2.34)

$$\eta_i = \overline{g}_{z_i^*,t_i}(g(\eta_i)). \tag{3.37}$$

Consequently, (with $g^0_{(\mathbf{z}^*,\mathbf{t})} \equiv 1_{z_0^*}$) Theorem 2.1 implies:

$$\begin{aligned} d_K(\eta_0, x^k) &\le \sum_{i=0}^{k-1} d_K(\overline{g}^i_{(\mathbf{z}^*,\mathbf{t})}(\eta_i), \overline{g}^{i+1}_{(\mathbf{z}^*,\mathbf{t})}(\eta_{i+1})) \\ &= \sum_{i=0}^{k-1} d_K(\overline{g}^{i+1}_{(\mathbf{z}^*,\mathbf{t})}(g(\eta_i)), \overline{g}^{i+1}_{(\mathbf{z}^*,\mathbf{t})}(\eta_{i+1})) \\ &\le \sum_{i=0}^{\infty}(1-(\theta/d))^{i+1}\epsilon \le (d/\theta)\epsilon. \end{aligned} \tag{3.38}$$

Taking the limit as $k \to \infty$ and applying (3.36) we obtain the $i = 0$ case of (3.31). The remaining cases follow from this one applied to the shifted chain $(s^+)^i(\eta)$ and to $(s^+_{\hat{G}})^i(\mathbf{z}^*, \mathbf{t})$.

(b) Apply part (a) to the one-sided projections of the shifted chain $(s)^i(\eta)$ and $(s_{\hat{G}})^i(\mathbf{z}^*, \mathbf{t})$ for all negative i. □

While the factor maps $\hat{h}$ and $\hat{h}^+$ and not one-to-one we can describe explicitly the preimages of points. First define

$$D\hat{K} \subset \hat{K} \times \hat{K} \times \hat{K}$$

$$(z_1^*, t_1, z_2^*, t_2, z_3^*, t_3) \in D\hat{K} \Leftrightarrow$$

$$z_2^* < z_1^* \cap z_3^* \text{ and } t_1(v) = t_2(v) = t_3(v) \text{ for all } v \in V(g(z_2^*)), \tag{3.39}$$

that is, $(z_2^*, t_2) < (z_1^*, t_1)$ and $(z_2^*, t_2) < (z_3^*, t_3)$, using the incidence relation of (2.33) for $\hat{K}$.

With $D\hat{G}$ on $D\hat{K}$ denoting the restriction of the product relation $\hat{G} \times \hat{G} \times \hat{G}$ on $\hat{K} \times \hat{K} \times \hat{K}$, let $s_{D\hat{G}}$ and $s^+_{D\hat{G}}$ be the shift maps on $D\hat{K}_{D\hat{G}}$ and $D\hat{K}^+_{D\hat{G}}$. Finally, let π_1, π_2 and π_3 denote the projections on the shift spaces to the corresponding shift spaces for $\hat{G}$ on $\hat{K}$.

4 Proposition. *Let $\hat{h} : \hat{K}_{\hat{G}} \to X_g$ and $\hat{h}^+ : \hat{K}^+_{\hat{G}} \to X$ be the shift factor maps associated with the simplicial dynamical system $g : K^* \to K$.*

$$\pi_1 \times \pi_3(D\hat{K}_{D\hat{G}}) = \hat{h}^{-1} \circ \hat{h} \subset \hat{K}_{\hat{G}} \times \hat{K}_{\hat{G}}$$

$$\pi_1 \times \pi_3(D\hat{K}^+_{D\hat{G}}) = (\hat{h}^+)^{-1} \circ \hat{h}^+ \subset \hat{K}^+_{\hat{G}} \times \hat{K}^+_{\hat{G}}. \tag{3.40}$$

Proof. Assume $\hat{h}(\mathbf{z}_1^*, \mathbf{t}_1) = \xi = \hat{h}(\mathbf{z}_3^*, \mathbf{t}_3)$. By (3.12)

$$\xi_i \in \langle z_{1i}^*, t_{1i} \rangle \cap \langle z_{3i}^*, t_{3i} \rangle \Rightarrow$$

$$\hat{q}(\xi_i) < (z_{1i}^*, t_{1i}) \text{ and } \hat{q}(\xi_i) < (z_{3i}^*, t_{3i}) \tag{3.41}$$

for all $i \in \mathbf{Z}$. Define $(\mathbf{z}_2^*, \mathbf{t}_2) = \hat{q}(\xi)$ to obtain a triple in $D\hat{K}_{D\hat{G}}$ which maps to $((\mathbf{z}_1^*, \mathbf{t}_1), (\mathbf{z}_3^*, \mathbf{t}_3))$ under $\pi_1 \times \pi_3$.

Conversely, if $(\mathbf{z}_1^*, \mathbf{t}_1, \mathbf{z}_2^*, \mathbf{t}_2, \mathbf{z}_3^*, \mathbf{t}_3)$ is an element of $D\hat{K}_{D\hat{G}}$ then let $\xi = h(\mathbf{z}_2^*, \mathbf{t}_2)$. As $\xi_i \in \langle z_{\alpha i}^*, t_{\alpha i} \rangle$ for all $i \in \mathbf{Z}$, $\alpha = 1, 2, 3$ and $\xi \in X_g$ it follows from the definition of the relation $\hat{h}$ and Theorem 2d that $\hat{h}(\mathbf{z}_\alpha^*, \mathbf{t}_\alpha) = \xi$ for $\alpha = 1, 2$. Hence, $((\mathbf{z}_1^*, \mathbf{t}_1), (\mathbf{z}_2^*, \mathbf{t}_2)) \in \hat{h}^{-1} \circ \hat{h}$.

The one-sided result has a completely similar proof. □

Let $g : K^* \to K$ be a simplicial map with $|K^*| = Y$ and $|K| = X$. A map $\gamma : Y \to X$ is called *subordinate* to g if

$$\gamma(z^*) \subset g(z^*) \text{ for all } z^* \in K^*. \tag{3.42}$$

If g is a simplicial dynamical system, so that K^* is a proper subdivision of K, $Y = X$, and $\hat{K}^*$ is defined by (3.3) then a map $\phi : X \to X$ is called *subordinate* to $1_{\hat{K}}$ if

$$\phi(\langle z^*, t \rangle) \subset \langle z^*, t \rangle \text{ for all } (z^*, t) \in \hat{K}. \tag{3.43}$$

For any set Y, we define on the set of maps from Y to X, the sup metric using d_K on X:

$$d_K(\gamma_1, \gamma_2) = \sup_{y \in Y} d_K(\gamma_1(y), \gamma_2(y)). \tag{3.44}$$

Using the isometry $c_K : X \to \mathbf{R}^{V(K)}$ we thus can regard the maps from Y to

X as a closed subset of the Banach space of maps from Y to $\mathbf{R}^{V(K)}$. Let M_g (and C_g) denote the set of maps (resp. continuous maps) subordinate to g and $M_{1_{\hat{K}}}$ (and $C_{1_{\hat{K}}}$) the set of maps (resp. continuous maps) subordinate to $1_{\hat{K}}$.

5 Lemma. *(a) The sets M_g. $M_{1_{\hat{K}}}$, C_g and $C_{g_{\hat{K}}}$ are closed convex subsets of the Banach space of maps from Y or X to $\mathbf{R}^{V(K)}$.*

$$\begin{array}{ccccc} M_{1_{\hat{K}}} & \subset & M_{1_{K^*}} & \subset & M_{1_K} \\ \cup & & \cup & & \cup \\ C_{1_{\hat{K}}} & \subset & C_{1_{K^*}} & \subset & C_{1_K} \end{array} \tag{3.45}$$

and the identity map $1_X \in C_{1_{\hat{K}}}$.

(b) For $\gamma \in C_g$, the inclusion in (3.42) is an equality for all $z^* \in K^*$. For $\phi \in C_{1_{\hat{K}}}$ the inclusion in (3.43) is an equality for all $(z^*, t) \in \hat{K}$.

Proof. Part (a) is obvious because the conditions (3.42) and (3.43) are preserved by convex combination and limits.

(b): By using induction on the dimension and restricting to a nondegenerate face if necessary we can assume that $\dim z^* = \dim g(z^*)$, so that g is a linear isomorphism, and that γ is surjective on the boundary faces. It suffices to prove that $g^{-1} \circ \gamma : z^* \to z^*$ hits every $x \in (z^*)^\circ$. If not then project away from it to obtain a continuous map from $z^* \to \partial z^*$ which agrees with $g^{-1} \circ \gamma$ on ∂z^*. By (a), γ is homotopic to g by maps in C_g and so the restriction of $g^{-1} \circ \gamma$ to ∂z^* is homotopic to the identity. Extending the homotopy we obtain a retraction from z^* to ∂z^*, which is impossible. Thus, we have equality in (3.42). The argument for (3.43) is similar. □

6 Theorem. *Let $g : K^* \to K$ be a simplicial dynamical system on $X = |K| = |K^*|$. For every map $\gamma : X \to X$ subordinate to g there is a unique map $\phi_\gamma : X \to X$ subordinate to $1_{\hat{K}}$ such that*

$$\phi_\gamma \circ \gamma = g \circ \phi_\gamma. \tag{3.46}$$

If d is the dimension of X and θ is the separation constant of K^* in K then

$$d(\phi_\gamma, 1_X) \le (d/\theta) d_K(\gamma, g). \tag{3.47}$$

Furthermore, if γ is continuous then ϕ_γ is continuous and so provides a semiconjugacy from γ on X to g on X.

Proof. Define the subset R of $X \times \hat{K}^+_{\hat{G}} \times X$ by

$$(x, (\mathbf{z}, \mathbf{t}), y) \in R \Leftrightarrow$$

(3.48) $$\gamma^i(x), g^i(y) \in \langle z_i^*, t_i \rangle \text{ for all } \mathbf{Z}_+.$$

Now for any $x \in X$ we use the carrier map $\hat{q} : X \to \hat{K}$ to define

(3.49) $$(z_i^*, t_i) = \hat{q}(\gamma^i(x)) \text{ for all } \mathbf{Z}_+.$$

By (3.42), we have $\gamma^{i+1}(x) \in \gamma(z_i^*) = g(z_i^*)$ and so by Lemma 2.5 $z_{i+1}^* = q(\gamma^{i+1}(x)) \subset g(z_i^*)$. That is, $z_{i+1}^* \in G^*(z_i^*)$ and so the sequence $\{(z_i^*, t_i)\}$ defines a point of $\hat{K}^+_{\hat{G}}$, which we will denote $(\mathbf{z}^*, \mathbf{t})_\gamma$. Now define

(3.50) $$\phi_\gamma(x) = \hat{h}^+((\mathbf{z}, \mathbf{t})_\gamma).$$

Clearly,

(3.51) $$(x, (\mathbf{z}, \mathbf{t})_\gamma, \phi_\gamma(x)) \in R.$$

Having defined the map ϕ_γ we observe that (3.49) with $i = 0$ implies that

(3.52) $$\phi_\gamma(x) \in \langle z_0, t_0 \rangle = \hat{q}(x)$$

which implies that ϕ_γ is subordinate to $1_{\hat{K}}$.

Now if $(x, (\mathbf{z}, \mathbf{t}), y) \in R$ then by (3.48), (3.49) and (3.12)

(3.53) $$(z_i^*, t_i)_\gamma = \hat{q}(\gamma^i(x)) < (z_i^*, t_i)$$

for all $i \in \mathbf{Z}_+$ and so by definition of the relation $\hat{h}^+$ and Proposition 4

(3.54) $$y = \hat{h}^+((\mathbf{z}, \mathbf{t})) = \hat{h}^+((\mathbf{z}, \mathbf{t})_\gamma) = \phi_\gamma(x).$$

Thus the projection of R to $X \times X$ is the mapping ϕ_γ. Clearly, $(x, (\mathbf{z}, \mathbf{t}), y) \in R$ implies $(\gamma(x), s^+_{\hat{G}}(\mathbf{z}, \mathbf{t}), g(y)) \in R$ and so

(3.55) $$\phi_\gamma(\gamma(x)) = g(\phi_\gamma(x)),$$

proving (3.46).

If $\epsilon = d_K(\gamma, g)$ then $\{\gamma^i(x) : i \in \mathbf{Z}_+\}$ is an ϵ chain for g on X and it is carried by $(\mathbf{z}, \mathbf{t})_\gamma$. So by Theorem 3

$$d_K(x, \phi_\gamma(x)) \leq (d/\theta)\epsilon, \tag{3.56}$$

proving (3.47).

Now if $\phi : X \to X$ is a map subordinate to $1_{\hat{K}}$ and $\phi(\gamma^i(x)) = g^i(\phi(x))$ for all $i \in \mathbf{Z}_+$ then

$$(x, (\mathbf{z}, \mathbf{t})_\gamma, \phi(x)) \in R \tag{3.57}$$

and so $\phi(x) = \phi_\gamma(x)$. This proves uniqueness of ϕ_γ.

Finally, if γ is continuous then the relation R is closed. By compactness, the relation ϕ_γ is a closed subset of $X \times X$ and so the map ϕ_γ is continuous.

Remark. There is an alternative proof of this result obtained by using an idea from Katok and Hasselblatt (1995, p. 77). On the complete space $M1_{\hat{K}}$ we define an operator $\phi \mapsto \phi'$ by letting ϕ' on $\langle z^*, t\rangle$ be defined to be $\overline{g}_{z*,t} \circ \phi \circ \gamma$. After showing ϕ' is well- defined we obtain a contraction map whose fixed point is ϕ_γ. This approach shows that the association $\gamma \mapsto \phi_\gamma$ is continuous. We will only need (3.47). □

There is a closely related result which is easiest to state when $g : K^* \to K$ is a nondegenerate simplicial dynamical system. When every simplex is nondegenerate $\hat{p} : \hat{K} \to K^*$ is a bijection and $\hat{p} : \hat{K}^+_{\hat{G}} \to K^{*+}_{G^*}$ is a homeomorphism identifying $s^+_{\hat{G}}$ with $s^+_{G^*}$.

7 Theorem. *Let $g : K^* \to K$ (with $X = |K| = |K^*|$) and $g_1 : K_1^* \to K_1$ (with $X_1 = |K_1| = |K_1^*|$) be simplicial dynamical systems with g_1 nondegenerate. Assume that $f : K^* \to K_1^*$ is a map of finite sets such that for every $z_0^*, z_1^* \in K^*$:*

$$z_0^* < z_1^* \Rightarrow f(z_0^*) < f(z_1^*); \tag{3.58}$$

and

$$z_1^* \subset g(z_0^*) \Rightarrow f(z_1^*) \subset g_1(f(z_0^*)). \tag{3.59}$$

There then exists a unique map $\phi : X \to X_1$ such that:

$$\phi(z^*) \subset f(z^*) \text{ for all } z^* \in K^* \tag{3.60}$$

and

$$\phi(g(x)) = g_1(\phi(x)) \text{ for all } x \in X. \tag{3.61}$$

Furthermore, ϕ is a continuous map.

Proof. Just as before, we define a closed subset R of $X \times \hat{K}^+_{\hat{G}} \times X_1$

$$(x, (\mathbf{z}, \mathbf{t}), y) \in R \Leftrightarrow$$

$$g^i(x) \in \langle z_i^*, t_i \rangle \text{ and } g_1^i(y) \in f(z_i^*) \text{ for all } i \in \mathbf{Z}_+. \tag{3.62}$$

The first condition says that $x = \hat{h}^+(\mathbf{z}, \mathbf{t})$. For the second, notice that condition (3.59) is equivalent to

$$(z_0^*, z_1^*) \in G^* \Rightarrow (f(z_0^*), f(z_1^*)) \in G_1^* \tag{3.63}$$

and so we can define $f(\mathbf{z}) \in K^+_{1G_1^*}$ by $f(\mathbf{z})_i = f(z_i)$ for all $i \in \mathbf{Z}_+$. The second condition in (3.62) says that $y = \hat{h}^+ \circ (\hat{p}^+)^{-1}(f(\mathbf{z}))$. So we obtain the map $\phi : X \to X_1$ by defining

$$\phi(x) = \hat{h}^+ \circ (\hat{p}^+)^{-1}(f(q^{*+}(x))). \tag{3.64}$$

Now proceed as in Theorem 6. In particular, $(x, (\mathbf{z}, \mathbf{t}), y) \in R$ implies $y = \phi(x)$ by using (3.47) and applying Proposition 4. □

Notice that the map $f : K^* \to K_1^*$ need not be a simplicial map. However, it might be. If K^* is a subdivision of K and K_1^* is a subdivision of K_1 then a pair of simplicial maps $f : K^* \to K_1^*$ and $f_0 : K \to K_1$ is called a *simplicial subdivision map* if the p.l. map associated with f is subordinate to the simplicial map f_0. This is equivalent to saying for $z^* \in K^*$ and $z \in K$:

$$z^* \subset z \Rightarrow f(z^*) \subset f_0(z). \tag{3.65}$$

8 Corollary. *With $X = |K| = |K^*|$ and $X_1 = |K_1| = |K_1^*|$, let $g : K^* \to K$ and $g_1 : K_1^* \to K_1$ be simplicial dynamical systems with g_1 nondegenerate. Assume that $f : K^* \to K_1^*$ and $f_0 : K \to K_1$ are a pair of simplicial maps forming a simplicial subdivision map. Assume, in addition, that*

$$f_0 \circ g = g_1 \circ f \tag{3.66}$$

as simplicial maps from K^* to K_1. There then exists a unique map $\phi : X \to X_1$ subordinate to f such that $\phi \circ g = g_1 \circ \phi$. Furthermore, ϕ is continuous.

Proof. The simplicial map f clearly satisfies (3.58). If $z_1^* \subset g(z_0^*)$ then by (3.65) and (3.66) $f(z_1^*) \subset f_0(g(z_0^*)) = g_1(f(z_0^*))$, i.e. (3.59) holds. Then (3.60) says that ϕ is subordinate to f and (3.61) says $\phi \circ g = g_1 \circ \phi$ on X. □

If K_1^* and K_2^* are subdivisions of K then a *subdivision isomorphism* (of subdivisions of K) is a simplicial isomorphism $r : K_1^* \to K_2^*$ whose associated p.l. map is subordinate to 1_K. That is, the pair $r : K_1^* \to K_2^*$ and $1_K : K \to K$ forms a simplicial subdivision map. The condition says that for each $z \in K$ the subcomplex of K_1^* which subdivides z is mapped by r to the subcomplex of K_2^* which subdivides z. K_1^* and K_2^* are called *isomorphic subdivisions* of K if such a subdivision isomorphism exists.

9 Proposition. *Let K_1^*, K_2^*, K_3^* be subdivisions of K and let $r_{21} : K_1^* \to K_2^*$ and $r_{32} : K_2^* \to K_3^*$ be simplicial isomorphisms.*

(a) r_{21} is a subdivision isomorphism iff its p.l. homeomorphism satisfies

$$r_{21}(z) = z \text{ for all } z \in K. \tag{3.67}$$

(b) If r_{21} is a subdivision isomorphism then $(r_{21})^{-1} : K_2^* \to K_1^*$ is a subdivision isomorphism. If r_{21} and r_{32} are subdivision isomorphisms then the composition $r_{32} \circ r_{21}$ is a subdivision isomorphism.

(c) If r_{21} and $\tilde{r}_{21} : K_1^* \to K_2^*$ are subdivision isomorphisms then $r_{21} = \tilde{r}_{21}$, i.e. there is at most one subdivision isomorphism between two subdivisions of K.

(d) If K_1^* and K_2^* are isomorphic subdivisions of K, then define

$$d_K(K_1^*, K_2^*) = d_K(r_{21}, 1_X) \tag{3.68}$$

where r_{21} is the p.l. homeomorphism associated with the unique subdivision isomorphism from K_1^* to K_2^*. This defines a metric on the set Iso($K^* : K$) of all subdivisions isomorphic to a given subdivision K^* of K.

(e) If K_1^* and K_2^* are isomorphic subdivisions and K_1^* is proper then K_2^* is proper.

Proof. (a) The definition requires $r_{21}(z) \subset z$ for all z, following (3.42). By continuity and Lemma 5b this is equivalent to (3.67).

(b) Since r_{21} is a bijection, (3.67) for r_{21} implies the same condition for $(r_{21})^{-1}$. Closure under composition is clear.

(c) By (b) it suffices to consider $r = (r_{21})^{-1} \circ \tilde{r}_{21} : K_1^* \to K_1^*$ and show that the identity is the only subdivision isomorphism from K_1^* to itself. Proceed by induction on $d = \dim K$. The initial step with $d = 0$ is obvious since $K_1^* = K$. Inductively, we can assume by applying the inductive hypotheses to the subdivision of the $d-1$ skeleton, that r is the identity on the $d-1$ skeleton of K. Then look at each d simplex z of K separately. This reduces to the case where $K = z$ and the subdivision isomorphism r on K_1^* is the identity on the boundary simplices.

Recall (see Appendix, Chapter 10) that a pair of simplices w_1^* and w_2^* of K_1^* *join* in K_1^* to form a simplex w^* of K_1^* if

$$V(w_1^*) \cap V(w_2^*) = \emptyset \text{ and } V(w^*) = V(w_1^*) \cup V(w_2^*). \tag{3.69}$$

We then write $w^* = w_1^* w_2^* = w_2^* w_1^*$. For $w_1^* \in K_1^*$ the *link* of w_1^*, denoted $\text{lk}(w_1^*; K_1^*)$ is the set of simplices which join with w_1^*, so that

$$\text{lk}(w_1^*; K_1^*) = \{w_2^* \in K_1^* : w_1^* w_2^* \in K^*\}. \tag{3.70}$$

For a d dimensional combinatorial manifold like the subdivision of z which we are considering, the link of a $d-1$ dimensional simplex w_1^* is a single point if w_1^* lies in the boundary $|\partial z|$ and is a pair of points if it does not. We also use the observation that if $r : K_1^* \to K_1^*$ is simplicial isomorphism then r restricts to an isomorphism of the link of w_1^* to the link of $r(w_1^*)$.

Given a vertex v of K_1^*, the subdivision of z, we can choose a sequence of distinct d dimensional simplices of $K_1^* : w_0^*, \ldots w_n^*$ such that (1) v is a vertex of w_n^*, (2) $w_i^* \cap w_{i-1}^* = z_i^*$ is a $d-1$ dimensional common face for $i = 1, \ldots, n$ and (3) $w_0^* \cap |\partial z| = z_0^*$ is a $d-1$ dimensional face in the boundary. For $i = 0, \ldots n$ let v_i be the unique vertex of $V(w_i^*) \backslash V(z_i^*)$. Thus, $w_i^* = z_i^* v_i$. Because the link of z_0^* in K_1^* is a single point it is v_0. But r is the identity on $z_0^* \subset |\partial z|$ and so r maps $\{v_0\} = \text{lk}(z_0^*, K_1^*)$ to itself. Thus, $r(v_0) = v_0$ and so r is the identity on w_0^*. In particular, r is the identity on the face z_1^* of w_0^*. So r maps $\text{lk}(z_1^*; K_1^*)$ to itself and this link consists of two points, namely $\text{lk}(z_1^*; w_0^*)$ and $\text{lk}(z_1^*; w_1^*) = \{v_1\}$. These are different points because $w_0^* \neq w_1^*$. Hence, $r(v_1) = v_1$ and so r is the identity on w_1^*. Proceed up the sequence, to obtain r the identity on w_n^* and hence on the original vertex v of K_1^* which we were examining.

(d) If f, g and h are maps on X and h is surjective then

$$d_K(f \circ h, g \circ h) = d_K(f, g). \tag{3.71}$$

So, in particular, with $r_{21} : K_1^* \to K_2^*$

$$d_K(r_{21}, 1_X) = d_K(r_{21} \circ (r_{21})^{-1}, (r_{21})^{-1}) = d_K(1_X, (r_{21})^{-1}) \tag{3.72}$$

proving $d_K(K_1^*, K_2^*) = d_K(K_2^*, K_1^*)$. The triangle inequality follows from

$$\begin{aligned} d_K(r_{31}, 1_X) &= d_K(r_{32} \circ r_{21}, 1_X) \leq \\ d_K(r_{32} \circ r_{21}, r_{21}) &+ d_K(r_{21}, 1_X) = \\ d_K(r_{32}, 1_X) &+ d_K(r_{21}, 1_X). \end{aligned} \tag{3.73}$$

Finally, $d_K(r_{21}, 1_X) = 0$ iff r_{21} is the identity map and so iff $K_1^* = K_2^*$.

(e) If $z_1^* \in K_1^*$ and $z \in K$ then $z_1^* \cap z \neq \emptyset$ iff $r_{21}(z_1^*) \cap z = r_{21}(z_1^* \cap z) \neq \emptyset$, by (3.67). Thus, z_1^* meets a pair of disjoint simplices of K iff $r_{21}(z_1^*)$ meets the same pair. □

From a simplicial dynamical system $g : K^* \to K$ and $K_1^* \in \text{Iso}(K^* : K)$ we obtain the simplicial dynamical system $g_1 : K_1^* \to K$ with

$$g_1 \circ r_1 = g \tag{3.74}$$

where r_1 is the subdivision isomorphism from K^* to K_1^*. We will call g_1 the K_1^* *isomorph* of g. Clearly, if r_{21} is the subdivision isomorphism from K_1^* to K_2^* then $r_{21} \circ r_1 = r_2$ implies

$$g_2 \circ r_{21} = g_1. \tag{3.75}$$

Notice that for $z_1^* \in K_1^*$ with $z_2^* = r_{21}(z_1^*)$, $g_1(z_1^*) = g_2(z_2^*)$, and for every vertex v of this common simplex of K

$$r_{21}((g_1|z_1^*)^{-1}(v)) = (g_2|z_2^*)^{-1}(v). \tag{3.76}$$

If $t_1 \in T^{z_1^*}$ then the composed map $t_2 \equiv r_{21} \circ t_1 \in T^{z_2^*}$ and as r_{21} maps the vertex set of $\langle z_1^*, t_1 \rangle$ to that of $\langle z_2^*, t_2 \rangle$, linearity implies:

$$r_{21}(\langle z_1^*, t_1 \rangle) = \langle z_2^*, t_2 \rangle$$

(3.77) where $z_2^* = r_{21}(z_1^*)$ and $t_2 = r_{21} \circ t_1$.

The equations (3.74) and (3.75) of simplicial maps hold for the associated p.l. maps as well. However, (3.74) does *not* say that the p.l. homeomorphism r_1 is a conjugacy between the dynamical systems on X induced by g_1 and g. We will now prove that such a topological conjugacy exists, but we will later see that it is usually not p.l.

10 Theorem. *Let $g : K^* \to K$ be a simplicial dynamical system on $X = |K| = |K^*|$ and let Iso$(K^* : K)$ be the space of subdivisions isomorphic to K^*. For $K_1^* \in \mathrm{Iso}(K^* : K)$ let $g_1 : K_1^* \to K$ be the K_1^* isomorph of g, so that $g_1 \circ r_1 = g$ where r_1 is the unique subdivision isomorphism from K^* to K_1^*. There is a unique homeomorphism $\rho_1 : X \to X$ such that*

$$\rho_1(\langle z^*, t\rangle) = r_1(\langle z^*, t\rangle) \text{ for all } (z^*, t) \in \hat{K} \tag{3.78}$$

and

$$g_1 \circ \rho_1 = \rho_1 \circ g. \tag{3.79}$$

Furthermore, if $K_2^* \in \mathrm{Iso}(K^* : K)$ with associated conjugating homeomorphism ρ_2, then

$$d_K(\rho_1, \rho_2) \le (1 + (d/\theta)) d_K(K_1^*, K_2^*) \tag{3.80}$$

where d is the dimension of X and θ is the maximum of the separation constants of K_1^* and K_2^* in K.

Proof. A subdivision isomorphism $r_{21} : K_1^* \to K_2^*$ is subordinate, as a p.l. map on X, to 1_K. Consequently, the p.l. map $r_{21} \circ g_2$ on X is subordinate to the simplicial map g_2. By Theorem 6 there is a unique continuous map ϕ_{21} on X which satisfies

$$\phi_{21}(\langle z_2^*, t_2\rangle) = \langle z_2^*, t_2\rangle \text{ for all } (z_2^*, t_2) \in \hat{K}_2 \tag{3.81}$$

and

$$\phi_{21} \circ r_{21} \circ g_2 = g_2 \circ \phi_{21}. \tag{3.82}$$

Now compose with r_{21} to define

$$\rho_{21} = \phi_{21} \circ r_{21}. \tag{3.83}$$

This is a continuous map on X and by (3.83) and (3.77) we have

$$\rho_{21}(\langle z_1^*, t_1\rangle) = r_{21}(\langle z_1^*, t_1\rangle) \text{ for all } (z_1^*, t_1) \in \hat{K}_1. \tag{3.84}$$

By (3.83) and (3.75) we have

$$\rho_{21} \circ g_1 = g_2 \circ \rho_{21}. \tag{3.85}$$

Conversely, if ρ_{12} satisfies (3.84) and (3.85), then $\phi_{21} = \rho_{21} \circ (r_{21})^{-1}$ satisfies (3.81) and (3.83) and so is the unique map given by Theorem 6. Thus, the continuous map ρ_{21} on X is characterized uniquely by conditions (3.78) and (3.79). In particular, if K_3^* is another element of $\mathrm{Iso}(K^* : K)$ then uniqueness implies:

$$\rho_{32} \circ \rho_{21} = \rho_{31}. \tag{3.86}$$

In particular, ρ_{21} is the inverse map for ρ_{21} and so each ρ_{21} is a homeomorphism providing a conjugacy from g_1 to g_2.

Now we apply (3.47) and (3.68) to get

$$\begin{gathered} d_K(\rho_{21}, 1_X) \leq \\ d_K(\phi_{21} \circ r_{21}, r_{21}) + d(r_{21}, 1_X) = \\ d_K(\phi_{21}, 1_X) + d(r_{21}, 1_X) \leq \\ (d/\theta_2) d_K(r_{21} \circ g_2, g_2) + d(r_{21}, 1_X) \leq \\ (1 + (d/\theta)) d_K(K_1^*, K_2^*) \end{gathered} \tag{3.87}$$

where θ is the separation constant of K_2^* in K. Since $d_K(\rho_{21}, 1_X) = d_K(\rho_{12}, 1_X)$, as in (3.72), we can replace θ by the separation constant of K_1^* in K, i.e. use the maximum.

Let ρ_α be the conjugacy homeomorphism from g to g_α for $\alpha = 1, 2$. By (3.86), $\rho_{21} \circ \rho_1 = \rho_2$ and so $d_K(\rho_2, \rho_1) = d_K(\rho_{21}, 1_X)$ yielding (3.80).

Remark. If θ_α is the separation constant for K_α^* ($\alpha = 1, 2$) in $\mathrm{Iso}(K^* : K)$ then it is easy to check that

$$|\theta_1 - \theta_2| \leq d_K(K_1^*, K_2^*). \tag{3.88}$$

□

Observe, finally, that if $K_1^* \in \mathrm{Iso}(K^* : K)$ with isomorphism $r_1 : K^* \to$

K_1^* and $g_1 \circ r_1 = g$ as in the theorem then the finite set map $r_1 \times 1_K : K^* \times K \to K_1^* \times K$ maps the relations g, J and j to the corresponding relations g_1, J_1 and j_1. Hence, we have

$$g \circ J^{-1} = G = g_1 \circ J_1^{-1} \text{ on } K$$

$$\begin{aligned} r_1 \times r_1(G^*) &= r_1 \times r_1(J^{-1} \circ g) \\ &= G_1^* = J_1^{-1} \circ g_1 \text{ on } K_1^*. \end{aligned} \tag{3.89}$$

Furthermore, if $t \in T^{z^*}$ then $t_1 = r \circ t \in T^{r(z^*)}$ satisfying (3.77), defines a map $\hat{r}_1 : \hat{K} \to \hat{K}_1$ such that

$$\hat{r}_1 \times \hat{r}_1(\hat{G}) = \hat{G}_1. \tag{3.90}$$

Hence, we obtain maps $r_1^+ : K_{G^*}^{*+} \to K_{1G_1^*}^{*+}$ and $\hat{r}_1 : \hat{K}_{\hat{G}}^+ \to \hat{K}_{1\hat{G}_1}^+$ associating the shifts and such that the following diagram commutes:

$$\begin{array}{ccc} K_{G^*}^{*+} & \xrightarrow{\;r_1^+\;} & K_{1G_1^*}^{*+} \\ \hat{p}^+ \uparrow & & \uparrow \hat{p}_1^+ \\ \hat{K}_{\hat{G}}^+ & \xrightarrow{\;\hat{r}_1^+\;} & \hat{K}_{1\hat{G}_1}^+ \\ \hat{h}^+ \downarrow & & \downarrow \hat{h}_1^+ \\ X & \xrightarrow[\;\rho_1\;]{} & X \end{array}$$

(3.91)

where ρ_1 is the homeomorphism satisfying (3.78) and (3.79).

The upper rectangle obviously commutes. For the lower let $(\mathbf{z}^*, \mathbf{t}) \in K_{\hat{G}}^+$ and $(\mathbf{w}^*, \mathbf{s}) \in \hat{K}_{1\hat{G}_1}^+$ with $(w_i^*, s_i) = \hat{r}_1(z_i^*, t_i)$ for all $i \in \mathbf{Z}_+$. Let $x \in X$ such that $g^i(x) \in \langle z_i^*, t_i \rangle$ for all $i \in \mathbf{Z}_+$. That is, $(\mathbf{w}^*, \mathbf{s}) = r_1^*(\mathbf{z}^*, \mathbf{t})$ and $x = \hat{h}^+(\mathbf{z}^*, \mathbf{t})$. By (3.79) and induction $\rho_1 \circ g^i = g_1^i \circ \rho_1$ for all $i \in \mathbf{Z}_+$. Hence, with $x_1 = \rho_1(x)$ we have

$$g_1^i(x_1) = \rho_1(g^i(x)) \in \rho_1\langle z_i^*, t_i \rangle = \langle w_i^*, s_i \rangle \tag{3.92}$$

for all $i \in \mathbf{Z}_+$. Hence, $x_1 = \hat{h}_1^+(\mathbf{w}^*, \mathbf{s})$ as required.

4. Recurrence and Basic Set Images

Because a simplicial map does not increase dimension on simplices, the dimension will be crucial for the study of recurrence.

1 Lemma. *Let K be a simplicial complex triangulating X with carrier mapping $q : X \to K$. Assume L is a set of k dimensional simplices of K, i.e. $L \subset S^k(K)$. For $x \in X$ the following conditions are equivalent:*

(1) $x \in |L| \backslash |\overline{S}^{k-1}(K)|$.

(2) $x \in |L|$ and $\dim q(x) \geq k$.

(3) $q(x) \in L$.

Proof. (1) $\Rightarrow$ (2): Since $x \in q(x)$, dim $q(x) < k$ implies $x \in |\overline{S}^{k-1}(K)|$.

(2) $\Rightarrow$ (3): Let $z \in L$ with $x \in z$ and so $q(x) < z$. Then $k = \dim z \geq \dim q(x) \geq k$ implies $q(x) = z$ by (2.18). So $q(x) \in L$.

(3) $\Rightarrow$ (1): With $z = q(x)$, $x \in (z)^\circ$ by definition. So $z \in L$ implies $x \in |L|$ and dim $z = k$ (all simplices of L have dimension k) implies $x \notin |\overline{S}^{k-1}(K)|$. □

Through the rest of this chapter $g : K^* \to K$ is a simplicial dynamical system on the polyhedron $X = |K| = |K^*|$. We study recurrence of the p.l. map $g : X \to X$ and its sample path shift homeomorphism $s_g : X_g \to X_g$ by using the factor maps $\hat{h} : \hat{K}_{\hat{G}} \to X_g$ and $\hat{h}^+ : \hat{K}^+_{\hat{G}} \to X$ described in the previous chapter.

If L is any nonempty subset of K^* then the restriction of G^* to L is $G^* \cap (L \times L)$. The associated sample path spaces satisfy

$$L_{G^*} = K^*_{G^*} \cap L^{\mathbf{Z}},$$

$$L^+_{G^*} = K^{*+}_{G^*} \cap L^{\mathbf{Z}_+} \tag{4.1}$$

These are just the sample path sequences for G^* each term of which lies in L.

Using $\hat{p} : \hat{K} \to K^*$ we define

$$\hat{L} = (\hat{p})^{-1}(L) = \cup\{\{z^*\} \times T^{z^*} : z^* \in L\}, \tag{4.2}$$

with the associated closed relation $\hat{G} \cap (\hat{L} \times \hat{L}) = (\hat{p} \times \hat{p})^{-1}(G \cap (L \times L))$. Via the maps $\hat{p} : \hat{K}_{\hat{G}} \to K^*_{G^*}$ and $\hat{p}^+ : \hat{K}^+_{\hat{G}} \to K^{*+}_{G^*}$ we have

$$\hat{L}_{\hat{G}} = \hat{K}_{\hat{G}} \cap \hat{L}^{\mathbf{Z}} = (\hat{p})^{-1}(L_{G^*})$$

$$\hat{L}^+_{\hat{G}} = \hat{K}^+_{\hat{G}} \cap \hat{L}^{\mathbf{Z}_+} = (\hat{p}^+)^{-1}(L^+_{G^*}). \tag{4.3}$$

We have defined the closed relations $\hat{h}^+ : \hat{K}^+_{\hat{G}} \to X$ and $\hat{h} : \hat{K}_{G^*} \to X_g$. We now define the closed relations

$$h = \hat{h} \circ (\hat{p})^{-1} : K^*_{G^*} \to X_g$$

$$h^+ = \hat{h}^+ \circ (\hat{p}^+)^{-1} : K^{*+}_{G^*} \to X. \tag{4.4}$$

For any subset L of K^* $h^+(L^+_{G^*}) = \hat{h}^+(\hat{L}^+_{\hat{G}})$ is a closed + invariant subset of X and $h(L_{G^*}) = \hat{h}(\hat{L}_{\hat{G}})$ is a closed invariant subset of X_g.

2 Lemma. *For $\xi \in X_g$, $\xi \in h(L_{G^*})$ iff $\xi_i \in h^+(L^+_{G^*})$ for all $i \in \mathbf{Z}$.*

Proof. $\xi = \hat{h}(\mathbf{z}, \mathbf{t})$ implies $\xi_i = \hat{h}^+(\pi^+_{\hat{G}}((s_{\hat{G}})^i(\mathbf{z}, \mathbf{t})))$. So $\xi \in h(L_{G^*})$ implies $\xi_i \in h^+(L^+_{G^*})$ for all i.

Conversely, suppose that for each $i \in \mathbf{Z}$ $(\mathbf{z}^*, \mathbf{t})^i \in \hat{L}^+_{\hat{G}}$ with $\xi_i = \hat{h}^+((\mathbf{z}^*, \mathbf{t})^i)$ and so $\xi_k \in \langle z^{*i}_{k-i}, t^i_{k-i} \rangle$ for all $k \geq i$. Extend the definition $(z^{*i}_j, t^i_j) \in \hat{K}$ arbitrarily for negative j and then let $(\mathbf{w}^*, \mathbf{s})^i = s^{-i}((\mathbf{z}^*, \mathbf{t})^i)$ obtaining a sequence in $\hat{K}^{\mathbf{Z}}$. Thus, for all $i \in \mathbf{Z}$ and all $k \geq i : \xi_k \in \langle w^{*i}_k, s^i_k \rangle, w^{*i}_k \in L$ and $w^{*i}_{k+1} \in G^*(w^{*i}_k)$. If $(\mathbf{z}^*, \mathbf{t})$ is a limit point in $\hat{K}^{\mathbf{Z}}$ of this sequence as $i \to -\infty$ then $(\mathbf{z}^*, \mathbf{t}) \in \hat{L}_{\hat{G}}$ and $\xi = \hat{h}(\mathbf{z}^*, \mathbf{t})$. □

As the notation in the above proof recalled, the closed relations $\hat{h}$ and $\hat{h}^+$ are functions because g is a simplicial dynamical system, i.e. K^* is a proper subdivision of K. On certain subsets, the relations h and h^+ are functions as well. Define:

$$ND(K^*) = \{z^* \in K^* : g \text{ is nondegenerate on } z^*\}. \tag{4.5}$$

$ND(K^*)$ is a subcomplex of K^*. If $L \subset ND(K^*)$ then $\hat{p} : \hat{L}_{\hat{G}} \to L_{G*}$ and $\hat{p}^+ : \hat{L}_{\hat{G}} \to L^+_{G^*}$ are homeomorphisms. So if $L \subset ND(K^*)$ then h and h^+ restrict to continuous mappings on the subsets L_{G^*} and $L^+_{G^*}$, respectively.

Recall that $G^* = J^{-1} \circ g$ on K^* and $G = g \circ J^{-1}$ on K where $J : K^* \to K$ is the inclusion relation containing the carrier map $j : K^* \to K$ (see (2.38)).

Now suppose $(z^*_1, z_1), (z^*_2, z_2) \in J$ with $g(z^*_1) = z_2$. By Lemma 2.6b and (2.22)

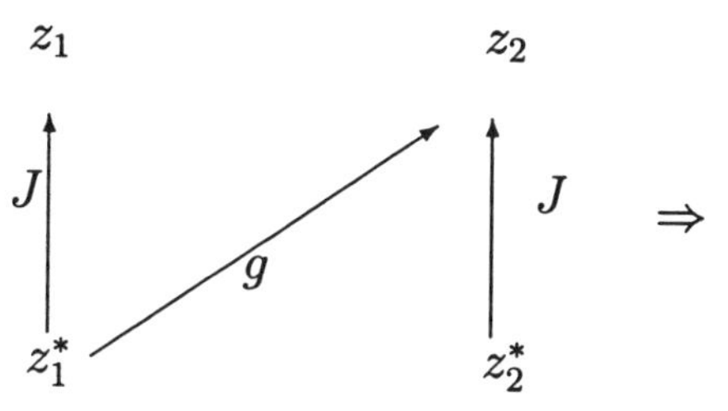

$$\dim z_1 \geq \dim z_1^* \geq \dim z_2 \geq \dim z_2^*. \tag{4.6}$$

In particular, we see again the inclusion (2.23): $g(\overline{S}^k(K^*)) \subset \overline{S}^k(K)$. It follows from (2.41) that the polyhedron $|\overline{S}^k(K^*)|$ and $|\overline{S}^k(K)|$ are closed subsets of X, + invariant for the p.l. map g.

3 Definition. *If $g : K^* \to K$ is a simplicial map with K^* a subdivision of K, then a simplex z^* of K^* is called an* expansion simplex *for g when*

$$\dim j(z^*) = \dim z^* = \dim g(z^*). \tag{4.7}$$

□

By (2.22) an expansion simplex is nondegenerate.

4 Proposition. *(a) Assume that $\{z_0^*, \ldots, z_m^*\}$ is a G^* chain in K^*. Let $z_0 = j(z_0^*)$ and $z_i = g(z_{i-1}^*)$ for $i = 1, \ldots, m$. If $\dim z_0 = \dim z_m^* = k$ then $\dim z_i^* = \dim z_i = k$ and $z_i = j(z_i^*)$ for $i = 0, \ldots, m$. In particular, z_i^* is an expansion simplex for $i = 0, \ldots, m-1$.*

(b) Assume that $\{z_0^*, \ldots, z_m^*\}$ is a G^* chain in K^*. Let $z_0 = j(z_0^*)$ and $z_i = g(z_{i-1}^*)$ for $i = 1, \ldots, m$. If $z_0^* = z_m^*$ and $\dim z_0^* = k$ then $\dim z_i^* = \dim z_i = k$, $z_i = j(z_i^*)$ and z_i^* is an expansion simplex for $i = 0, \ldots, m$.

(c) The cyclic set $|\mathcal{O}G^*| \subset K^*$ consists entirely of expansion simplices. If B^* is a basic set for G^*, i.e. an $\mathcal{O}G^* \cap \mathcal{O}G^{*-1}$ equivalence class in $|\mathcal{O}G^*|$, then the dimension function is constant on B^*. We call B^* a k skeleton basic set *for G^* when this common dimension is k. For z_1^*, z_2^* in B^**

$$z_2^* \in G^*(z_1^*) \Rightarrow g(z_1^*) = j(z_2^*). \tag{4.8}$$

(d) If B^* is a k skeleton basic set for G^* then there is a unique G basic set B consisting of k simplices of K, called the k skeleton basic set for G associated with B^*, *which is defined by the equations*

$$g(B^*) = B = j(B^*). \tag{4.9}$$

Conversely, if B is a G basic set then it is associated with a unique G^* basic set B^* obtained by the equation

$$B^* = \{z^* \in L^* : g(z^*), j(z^*) \in B\}. \tag{4.10}$$

(e) If z^* is an expansion simplex of dimension k then $j(z^*) \in S^k(K)$. If B^* is a k skeleton basic set for G^* with associated G basic set B then

$$B^* \subset j^{-1}(B) \subset j^{-1}(S^k(K)) \subset J^{-1}(\overline{S}^k(K)). \tag{4.11}$$

(f) Because $g(V(K^*)) \subset V(K)$ and $V(K) \subset V(K^*)$, g restricts to a mapping of $V(K)$ into itself. On the nonempty set $\cap_{n=0}^{\infty} g^n(V(K))$, g restricts to a permutation, and so the set consists of periodic points for g. A 0 skeleton basic set for G^* (= the associated 0 skeleton basic set for G) is exactly a periodic orbit of the map g on the set of vertices $V(K)$.

Proof. (a) By (4.6), $\dim z_i \geq \dim z_{i+1} \geq \dim z^*_{i+1}$ for $i = 0, \ldots, m-1$. So if $\dim z_0 = \dim z^*_m = k$ all of the intervening simplices have dimension k as well. As $z_{i+1} \in J(z^*_{i+1})$, Lemma 2.6b implies $z_{i+1} = j(z^*_{i+1})$ for $i = 0, \ldots, m-1$. Hence, $z^*_0, \ldots, z^*_{m-1}$ are expansion simplices.

(b) As in (a), $k = \dim z^*_0 \geq \dim z_m \geq \dim j(z^*_m) \geq \dim z^*_m$. Because $z^*_0 = z^*_m$ these are all equalities. In particular, $z_0 = j(z^*_m)$ and $z^*_m = z^*_0$ have dimension k. The conclusion follows from (a). Because $z^*_m = z^*_0$ it, too, is an expansion simplex.

(c) Given w^*_1 and w^*_2 in the same $\mathcal{O}G^* \cap \mathcal{O}G^{*-1}$ class of $|\mathcal{O}G^*|$ (including the case $w^*_1 = w^*_2$ any element of $|\mathcal{O}G^*|$) there is some G^* chain $\{z^*_0, \ldots, z^*_m\}$ with $z^*_0 = z^*_m$ and including w^*_1 and w^*_2. So by (b) they are all expansion simplices of the same dimension. If $z^*_2 \in G^*(z^*_1)$ with z^*_1, z^*_2 in such a basic set then $\dim z^*_1 = \dim z^*_2$ implies $g(z^*_1) = j(z^*_2)$ by Lemma 2.6b.

(d) Proposition 1.4 describes the bijective association between a G^* basic sets and G basic sets. The association between B^* and B is given by $B = g(B^*)$ and $B^* = g^{-1}(B) \cap J^{-1}(B)$, see (1.26). In particular, if B^* is a k skeleton basic set for G^* then (c) implies that $B = g(B^*)$ consists of

K simplices of dimension k. If $z \in B$ and $z^* \in B^*$ with $z \in J(z^*)$ then $\dim z = \dim z^* = k$ and so Lemma 2.6b implies $z = j(z^*)$. Thus,

$$B^* = g^{-1}(B) \cap j^{-1}(B) \tag{4.12}$$

which restates (4.10). In particular, $j(B^*) \subset B$. On the other hand, $z \in B$ implies $z = g(z_1^*)$ for some $z_1^* \in B^*$. Because B^* is an $\mathcal{O}G^* \cap \mathcal{O}G^{*-1}$ equivalence class there exists $z_2^* \in B^*$ such that $z_2^* \in G^*(z_1^*)$, i.e. $z = g(z_1^*) \in J(z_2^*)$. As before $z = j(z_2^*)$ and so $B \subset j(B^*)$. Together with (1.26) this proves (4.9).

(e) The first sentence is clear. (4.11) follows from part (d).

(f) If $v \in V(K)$, is a vertex of K, then $J^{-1}(v)$ is the singleton set consisting of this same v regarded as a vertex of $V(K^*)$. That is, J^{-1} on $V(K)$ is the inclusion mapping of $V(K)$ into $V(K^*)$. Hence, G^* on $V(K^*)$ is the map $g : V(K^*) \to V(K)$ and G is the restriction of this map to the subset $V(K)$. As these relations are maps the rest follows from standard results on maps of finite sets, i.e. a surjective endomorphism is a permutation and so is a product of disjoint cycles. □

In particular, every periodic simplex in K^* is an expansion simplex and so is nondegenerate, i.e.

$$|\mathcal{O}G^*| \subset ND(K^*). \tag{4.13}$$

Furthermore, while not every k dimensional simplex need be nondegenerate any G^* chain consisting entirely of k dimensional simplices contains only nondegenerates. Thus,

$$\begin{aligned} &S^k(K^*)_{G^*} \subset ND(K^*)_{G^*} \\ &S^k(K^*)^+_{G^*} \subset ND(K^*)^+_{G^*}. \end{aligned} \tag{4.14}$$

In particular, on these sets the relations h and h^+ are mappings.

In the language of Chapter 1, condition (4.13) says that the fat shift $\hat{G}$ associated with G^* is only finitely degenerate. So $\hat{p} : \hat{K}_{\hat{G}} \to K^*_{G^*}$ and $\hat{p}^+ : \hat{K}^+_{\hat{G}} \to K^{*+}_{G^*}$ are each injective over chain recurrent points. The fibre over any point is easy to describe.

Clearly, $z^* \in K^*$ implies

$$\dim g(z^*) + \dim T^{z^*} = \dim z^*. \tag{4.15}$$

Hence, if $\{z_0^*, \ldots, z_m^*\}$ is a G^* chain then (4.6) implies

$$\sum_{i=0}^{m-1} \dim T^{z_i^*} \leq \dim z_0^* - \dim z_m^* \leq \dim X. \tag{4.16}$$

It follows that any fibre for $\hat{p}$ or $\hat{p}^+$ is a closed, convex cell of dimension at most that of X.

For B^* a G^* basic set we define the sample path spaces following (4.1)

$$\begin{aligned} B^*_{G^*} &= K^*_{G^*} \cap B^{*\mathbf{Z}} \\ B^{*+}_{G^*} &= K^{*+}_{G^*} \cap B^{*\mathbf{Z}_+}. \end{aligned} \tag{4.17}$$

By Proposition 1.1 these are the basic sets for s_{G^*} and $s^+_{G^*}$, respectively. A basic set consists of expansion simplices and so h and h^+ are continuous maps on these subsets. The closed, invariant subsets $h(B^*_{G^*}) \subset X_g$ and $h^+(B^{*+}_{G^*}) \subset X$ are called *basic set images.* These basic set images will be our primary tool for the study of recurrence for the dynamical systems g on X and s_g on X_g.

As G^* and hence s_{G^*} has only finitely many basic sets the collection $\{B^*_{G^*} : B^* \text{ a basic set for } G^*\}$ is a fine decomposition for the homeomorphism s_{G^*}, see Propositions 1.4 and 1.5. It follows from Proposition 1.6 that $\{h(B^*_{G^*}) : B^* \text{ a basic set for } G^*\}$ forms a fine predecomposition for the homeomorphism s_g. However, it will be useful to prove this directly.

5 Proposition. *(a) For any $\mathbf{z}^* \in K^{*+}_{G^*}$ there is a unique basic set for G^* denoted $B^*(\mathbf{z}^*)$, such that $z_i^* \in B^*(\mathbf{z}^*)$ for sufficiently large i. $B^*(\mathbf{z}^*)$ is called the* endset *for $\mathbf{z}^*$. For any $\mathbf{z}^* \in K^*_{G^*}$ there are unique basic sets for G^*, denoted $B^*(\mathbf{z}^*)$ and $A^*(\mathbf{z}^*)$, such that $z_i^* \in B^*(\mathbf{z}^*)$ and $z_{-i}^* \in A^*(\mathbf{z}^*)$ for sufficiently large i. $B^*(\mathbf{z}^*)$ and $A^*(\mathbf{z}^*)$ are called the* endset *and* startset *for $\mathbf{z}^*$, respectively.*

(b) For any $x \in X$ the endset of $q^{*+}(x) \in K^{*+}_{G^*}$ is denoted $B^*(x)$ and called the endset of x. Thus, $B^*(x)$ is the unique G^* basic set such that the carrier simplices $q^*(g^i(x))$ in K^* lie in $B^*(x)$ for sufficiently large i. For any $\xi \in X_g$ the endset and startset of $q^*(\xi) \in K^*_{G^*}$ are denoted $B^*(\xi)$ and $A^*(\xi)$. These are the unique basic sets such that $q^*(\xi_i) \in B^*(x)$ and $q^*(\xi_{-i}) \in A^*(x)$ for sufficiently large i.

Proof. For $\mathbf{z}^* \in K^{*+}_{G^*}$ we have that z_i^* eventually lies in some G^* basic set by Lemma 1.3. Since distinct basic sets are disjoint the endset $B^*(\mathbf{z}^*)$ is thus uniquely defined. Applying the same argument to the inverse relation $(G^*)^{-1}$ we obtain the startset $A^*(\mathbf{z}^*)$ for $\mathbf{z}^* \in K^*_{G^*}$. This proves (a) from which (b) clearly follows. □

Since the startsets and endsets depend only on orbit entries near infinity we clearly have, for $x \in X$ and $\xi \in X_g$:

$$B^*(x) = B^*(g^i(x)) \text{ for all } i \in \mathbf{Z}_+$$

$$B^*(\xi) = B^*(s_g^i(\xi)) = B^*(\xi_i) \text{ for all } i \in \mathbf{Z}$$

$$A^*(\xi) = A^*(s_g^i(\xi)) \text{ for all } i \in \mathbf{Z}. \tag{4.18}$$

6 Proposition. *(a) Let $x \in X$ and B^* be G^* basic sets. Assume there exists $(\mathbf{z}^*, \mathbf{t}) \in \hat{K}^+_{\hat{G}}$ such that $x = \hat{h}^+(\mathbf{z}^*, \mathbf{t})$ and such that with $\mathbf{z}^* = \hat{p}^+(\mathbf{z}^*, \mathbf{t})$, B^* is the endset of $\mathbf{z}^*$. Then for sufficiently large $i \in \mathbf{Z}_+$ $g^i(x) \in h^+(B^{*+}_{G^*})$ and so $\omega g(x) \subset h^+(B^{*+}_{G^*})$. In particular, if B^* is the endset for x then $g^i(x) \in h^+(B^{*+}_{G^*})$ for sufficiently large i.*

(b) Let $\xi \in X_g$ and A^*, B^* be G^* basic sets. Assume there exists $(\mathbf{z}^*, \mathbf{t}) \in \hat{K}_{\hat{G}}$ such that $\xi = \hat{h}(\mathbf{z}^*, \mathbf{t})$ and such that with $\mathbf{z}^* = \hat{p}(\mathbf{z}^*, \mathbf{t})$, A^* and B^* are, respectively, the startset and endset of $\mathbf{z}^*$. Then there exist $\zeta \in h(A^*_{G^*})$ and $\eta \in h(B^*_{G^*})$ such that

$$\mathrm{Lim}_{i \to \infty} d_K(\xi_{-i}, \zeta_{-i}) = 0$$

$$\mathrm{Lim}_{i \to \infty} d_K(\xi_i, \eta_i) = 0, \tag{4.19}$$

from which it follows that $\alpha s_g(\xi) \subset h(A^*_{G^*})$, and $\omega s_g(\xi) \subset h(B^*_{G^*})$. In particular these results hold if A^* and B^* are, respectively, the startset and endset for ξ.

Proof. (a) For any $i \in \mathbf{Z}_+$, (3.18) implies

$$g^i(x) = \hat{h}^+((s^+_{\hat{G}})^i(\mathbf{z}^*, \mathbf{t})). \tag{4.20}$$

By definition of the endset there exists $i_0 \in \mathbf{Z}_+$ such that $z^*_i \in B^*$ for all $i \geq i_0$. Hence, for $i \geq i_0$, $(s^+_{G^*})^i(\mathbf{z}^*) \in B^{*+}_{G^*}$ on which h^+ is a map so that

$$\hat{h}^+((s^+_{\hat{G}})^i(\mathbf{z}^*, \mathbf{t})) = h^+((s^+_{G^*})^i(\mathbf{z}^*)) \tag{4.21}$$

and so by (4.20), $g^i(x) \in h^+(B^{*+}_{G^*})$ for $i \geq i_0$. As $h^+(B^{*+}_{G^*})$ is closed the limit point set $\omega g(x)$ is contained in it.

For any x, $\hat{h}^+(\hat{q}^+(x)) = x$ and $\hat{p}^+(\hat{q}^+(x)) = q^{*+}(x)$ so that the hypothesis holds with $(\mathbf{z}^*, \mathbf{t}) = \hat{q}^+(x)$ if B^* is the endset of x.

(b) There exists $i_0 \in \mathbf{Z}_+$ such that $z_i^* \in B^*$ and $z_{-i}^* \in A^*$ for all $i \geq i_0$. Because basic sets are surjective subsets we can construct, inductively, $\mathbf{w}^* \in B^*_{G^*}$ and $\mathbf{u}^* \in A^*_{G^*}$ such that

$$z_i^* = w_i^* \text{ and } z_{-i}^* = u_{-i}^* \text{ for } i \geq i_0. \tag{4.22}$$

Let $\zeta = h(\mathbf{u}^*) \in h(A^*_{G^*})$ and $\eta = h(\mathbf{w}^*) \in h(B^*_{G^*})$.

Notice first that for $i \geq i_0$

$$\pi^+_{G^*}((s_{G^*})^i(\mathbf{z}^*)) = \pi^+_{G^*}((s_{G^*})^i(\mathbf{w}^*)) \tag{4.23}$$

where $\pi^+_{G^*}$ is the projection from $K^*_{G^*}$ to $K^{*+}_{G^*}$. By (3.18) it follows that

$$\xi_i = \eta_i \quad \text{for} \quad i \geq i_0. \tag{4.24}$$

A fortiori, $\text{Lim } d_K(\xi_i, \eta_i) = 0$ as $i \to \infty$.

On the other hand, for $i > i_0$ we have ξ_{-i} and $\zeta_{-i} \in z_{-i}^*$ and so by (2.24)

$$\begin{aligned} \xi_{-i} &= \bar{g}_{z_{-i}^*}(\xi_{-i+1}) \\ \zeta_{-i} &= \bar{g}_{z_{-i}^*}(\zeta_{-i+1}). \end{aligned} \tag{4.25}$$

By Theorem 2.11 these local inverses contract distance, and so $d_K(\xi_{-i}, \zeta_{-i}) \to 0$ as $i \to \infty$.

We have shown that the sequences $\{\xi_i\}$ and $\{\eta_i\}$ are asymptotic in X as $i \to \infty$. It follows that the sequences $\{(s_g)^i(\xi)\}$ and $\{(s_g)^i(\eta)\}$ are asymptotic in X_g as $i \to \infty$. Hence, they have the same limit point set, i.e. $\omega s_g(\xi) = \omega s_g(\eta) \subset h(B^*_{G^*})$. Similarly, $\alpha s_g(\xi) = \alpha s_g(\zeta) \subset h(A^*_{G^*})$.

Finally, as in (a), we can use $(\mathbf{z}^*, \mathbf{t}) = \hat{q}(\xi) \in \hat{K}_{\hat{G}}$ with $\mathbf{z}^* = q^*(\xi) \in K^*_{G^*}$, when A^* and B^* are the startset and endset for ξ. □

Since we will want to concatenate the basic set images (see (1.33)), we want to be able to recognize when two of them intersect.

By analogy with (3.39) define

$$DK^* = \{(z_1^*, z_2^*, z_3^*) \in K^* \times K^* \times K^* : z_2^* < z_1^* \cap z_3^*\} \tag{4.26}$$

and let DG^* denote the restriction to DK^* of the product relation $G^* \times G^* \times G^*$, i.e.

$$DG^* = (G^* \times G^* \times G^*) \cap (DK^* \times DK^*). \tag{4.27}$$

7 Proposition. *(a) Let B^* and A^* be basic sets for G^*. The following conditions are equivalent:*

(1) $h(B^*_{G^*}) \cap h(A^*_{G^*}) \neq \emptyset$.

(2) $h^+(B^{*+}_{G^*}) \cap h^+(A^{*+}_{G^*}) \neq \emptyset$.

(3) There exists C^* a basic set for DG^* on DK^* such that $\pi_1(C^*) \subset B^*$ and $\pi_3(C^*) \subset A^*$.

(4) There exist a positive integer m, a point $x \in X$ and G^* chains $\{z^*_0, \ldots, z^*_m\}$ in B^*, $\{w^*_0, \ldots, w^*_m\}$ in A^* satisfying

$$z^*_0 = z^*_m, \ w^*_0 = w^*_m, \ x = g^m(x)$$

$$g^i(x) \in z^*_i \cap w^*_i \text{ for } i = 0, \ldots, m. \tag{4.28}$$

Furthermore, $h(B^*_{G^*}) \subset h(A^*_{G^*})$ iff $h^+(B^{*+}_{G^*}) \subset h^+(A^{*+}_{G^*})$.

(b) Let B^* and A^* be k skeleton basic sets for G^* with B and A the respective associated k skeleton basic sets for G. If $B^* \neq A^*$,

$$h^+(B^{*+}_{G^*}) \cap h^+(A^{*+}_{G^*}) \subset |B^*| \cap |A^*| \subset |B| \cap |A| \subset |S^{k-1}(K)|. \tag{4.29}$$

(c) Let B^* be a basic set for G^* and $\mathcal{B}$ a collection of basic sets for G^*. If for every periodic point x in $h^+(B^{*+}_{G^*})$ there exists $A^* \in \mathcal{B}$ such that $x \in h^+(A^*_{G^*})$ then there exists $A^* \in \mathcal{B}$ such that $h^+(B^{*+}_{G^*}) \subset h^+(A^{*+}_{G^*})$.

Proof. (a) For $\pi^+_{G^*} : K^*_{G^*} \to K^{*+}_{G^*}$ we have $\pi^+_{G^*}(B^*_{G^*}) = B^{*+}_{G^*}$ by (1.19). Now apply $\pi_0 : X_g \to X$ to get:

$$\pi_0(h(B^*_{G^*})) = h^+(\pi^+_{G^*}(B^*_{G^*})) = h^+(B^{*+}_{G^*}). \tag{4.30}$$

So (1) $\Rightarrow$ (2) and $h(B^*_G) \subset (A^*_{G^*})$ implies $h^+(B^{*+}_{G^*} \subset h^+(A^{*+}_{G^*})$.

On the other hand, if $h^+(B^{*+}_{G^*}) \subset h^+(A^{*+}_{G^*})$ then it follows from Lemma 2 that for $\xi \in h(B^*_{G^*})$, $\xi_i \in h^+(B^{*+}_{G^*})$ and hence $\xi_i \in h^+(A^{*+}_{G^*})$ for all i. By Lemma 2 again $\xi \in h(A^*_{G^*})$. Hence, $h(B^*_{G^*}) \subset h(A^*_{G^*})$.

(2) $\Rightarrow$ (3): If $h^+(\mathbf{z}^*) = h^+(\mathbf{w}^*)$ with $\mathbf{z}^* \in B^{*+}_{G^*}$ and $\mathbf{w}^* \in A^{*+}_{G^*}$ then since

$\hat{p}^+$ is injective over $B^{*+}_{G^*}$ and $A^{*+}_{G^*}$ it follows from Proposition 3.4 that there exists $(\mathbf{z}^*_1, \mathbf{z}^*_2, \mathbf{z}^*_3) \in (DK^*)_{DG^*}$ with $\mathbf{z}^*_1 = \mathbf{z}^*$ and $\mathbf{z}^*_3 = \mathbf{w}^*$. By Lemma 1.3, there is a DG^* basic set C^* in DK^* which is the endset for $(\mathbf{z}^*_1, \mathbf{z}^*_2, \mathbf{z}^*_3)$. The projection $\pi_1 : DK^* \to K^*$ maps DG^* to G^* and so maps the basic set C^* into some G^* basic set $\tilde{B}^*$. Because $z^*_i \in B^*$ for all i and $z^*_i \in \tilde{B}^*$ for large enough i, it follows that $B^* = \tilde{B}^*$ since distinct basic sets are disjoint. Similarly, $\pi_3(C^*) \subset A^*$.

(3) $\Rightarrow$ (4): Because the basic set C^* for DG^* is an $\mathcal{O}(DG^*) \cap \mathcal{O}(DG^*)^{-1}$ equivalence class, we can start with a periodic DG^* chain and extend to get $(\mathbf{z}^*_1, \mathbf{z}^*_2, \mathbf{z}^*_3) \in C^*_{DG^*}$ which is a periodic sequence of period m, i.e. $z^*_{\alpha(i+m)} = z^*_{\alpha_i}$ for $\alpha = 1, 2, 3$ and all $i \in \mathbf{Z}$. In particular, these are all expansion simplices for G^* and so $\xi = h(\mathbf{z}^*_2)$ is the unique element of X_g such that $\xi_i \in z^*_{2i}$ for all i. Hence, by uniqueness, $\xi_{i+m} = \xi_i$ for all i. Thus, letting $z^*_i = z^*_{1i}$, $w^*_i = z^*_{3i}$ for $i = 0, \dots, m$ and $x = \xi_0$ we obtain (4.28).

(4) $\Rightarrow$ (1): Let $\mathbf{z}^*$ and $\mathbf{w}^*$ be the periodic extension of the chains $\{z_0, \dots, z_m\}$ and $\{w^*_0, \dots w^*_m\}$ to elements of $B^*_{G^*}$ and $A^*_{G^*}$, respectively. Let ξ be the periodic extension of the sequence $\{x, \dots, g^m(x)\}$ to a sequence in X_g. Clearly, $\xi_i \in z^*_i \cap w^*_i$ for all $i \in \mathbf{Z}$ and so $h(\mathbf{z}^*) = \xi = h(\mathbf{w}^*)$, i.e. ξ is a point in $h(B_{G^*}) \cap h(A_{G^*})$.

(b) Since $B = j(B^*)$ by (4.9) we have $h^+(B^{*+}_{G^*}) \subset |B^*| \subset |B|$ which implies the first two inclusions of (4.29). If $x \in |B| \backslash |\overline{S}^{k-1}(K)|$ then by Lemma 1 we have $q(x) \in B$. So if $x \in (|B| \cap |A|) \backslash |\overline{S}^{k-1}(K)|$ then $q(x) \in B \cap A$. As distinct basic sets for G are disjoint $B = A$. By (4.10), $B^* = A^*$.

(c) On the basic set $B^{*+}_{G^*}$ the shift of finite type $s^+_{G^*}$ restricts to a topologically transitive map with dense periodic points. So by Taylor's Lemma 1.2 we can choose a sequence $\{\mathbf{z}^{*n}\}$ of periodic points in $B^{*+}_{G^*}$ such that the orbit of $\mathbf{z}^{*n}$ is $1/n$ dense in $B^{*+}_{G^*}$. By hypothesis, we can choose for each n, $A^{*n} \in \mathcal{B}$ such that $h^+(\mathbf{z}^{*n}) \in h^+(A^{*n+}_{G^*})$. Because $\mathcal{B}$ is a finite set there exists $A^* \in \mathcal{B}$ such that $h^+(\mathbf{z}^{*n'}) \in h^+(A^{*+}_{G^*})$ for some subsequence $\{\mathbf{z}^{*n'}\}$. But every point of $h^+(B^{*+}_{G^*})$ is a limit of some sequence of points chosen on the orbits of the points $\{h^+(\mathbf{z}^{*n'})\}$. Thus, every point of $h^+(B^{*+}_{G^*})$ is contained in the closed invariant set $h^+(A^{*+}_{G^*})$. □

For a collection $\mathcal{B}$ of G^* basic sets, we say that $B^* \in \mathcal{B}$ has a $\mathcal{B}$ *maximal image* or a *maximal image with respect to* $\mathcal{B}$ if it satisfies the equivalent implications

$$\begin{array}{ll} & \tilde{B}^* \in \mathcal{B} \ \text{ and } \ h^+(B^{*+}_{G^*}) \subset h^+(\tilde{B}^{*+}_{G^*}) \Rightarrow h^+(B^{*+}_{G^*}) = h^+(\tilde{B}^{*+}_{G^*}) \\ (4.31) & \tilde{B}^* \in \mathcal{B} \ \text{ and } \ h(B^*_{G^*}) \subset h(\tilde{B}^*_{G^*}) \Rightarrow h(B^*_{G^*}) = h(\tilde{B}^*_{G^*}). \end{array}$$

A basic set B^* has a *maximal image* if its image is maximal with respect to the entire set of basic sets of G^*. The *natural pre-decomposition* $\mathcal{F}_g$ for X_g is

$$(4.32) \qquad \mathcal{F}_g = \{h(B^*_{G^*}) : B^* \text{ is a basic set with maximal image}\}.$$

8 Proposition. *For a collection $\mathcal{B}$ of G^* basic sets the following are equivalent.*

(1) $h^+(|\mathcal{C}s^+_{G^*}|) \subset \cup\{h^+(B^{*+}_{G^*}) : B^* \in \mathcal{B}\}$.

(2) For every periodic point x in X there exists $B^* \in \mathcal{B}$ such that $x \in h^+(B^{*+}_{G^*})$.

(3) For every basic set $\tilde{B}^*$ with maximal image there exists $B^* \in \mathcal{B}$ such that $h^+(B^{*+}_{G^*}) = h^+(\tilde{B}^{*+}_{G^*})$.

When these conditions hold then $\mathcal{F}_g = \{h(B^*_{G^*}) : B^* \in \mathcal{B}$ has a $\mathcal{B}$ maximal image$\}$.

Proof. (1) $\Rightarrow$ (2): Clearly, $q^{*+}(x) \in K^{*+}_{G^*}$ is $s^+_{G^*}$ periodic if x is a g periodic point. In that case, $q^{*+}(x) \in |\mathcal{C}s^+_{G^*}|$ and $x = h^+(q^{*+}(x)) \in h^+(|\mathcal{C}s^+_{G^*}|)$.

(2) $\Rightarrow$ (3): Every periodic point in $h^+(\tilde{B}^{*+}_{G^*})$ is contained in some $\mathcal{B}$ basic set image by assumption (2). So by Proposition 7a, there exists $B^* \in \mathcal{B}$ such that $h^+(\tilde{B}^{*+}_{G^*}) \subset h^+(B^{*+}_{G^*})$. But $\tilde{B}^*$ is assumed to have a maximal image. So the inclusion must be an equality.

(3) $\Rightarrow$ (1): If $\mathbf{z}^* \in |\mathcal{C}s^+_{G^*}|$ then $h^+(\mathbf{z}^*)$ is in some basic set image. Let $h^+(\tilde{B}^{*+}_{G^*})$ be maximal among the basic set images containing $h^+(\mathbf{z}^*)$. $\tilde{B}^*$ is a basic set with maximal image and so by (3), $h^+(\tilde{B}^{*+}_{G^*}) = h^+(B^{*+}_{G^*})$ for some $B^* \in \mathcal{B}$. Thus, $h^+(\mathbf{z}^*) \in h^+(B^{*+}_{G^*})$ and (1) follows.

Now assume (3). If $F \in \mathcal{F}_g$ then $F = h(\tilde{B}^*_{G^*})$ for some basic set $\tilde{B}^*$ with maximal image. By (3) and Proposition 7a, $h(\tilde{B}^*_{G^*}) = h(B^*_{G^*})$ for some B^* in $\mathcal{B}$. Since $\tilde{B}^*$ has maximal image so does B^* and a fortiori it is $\mathcal{B}$ maximal. On the other hand, if $F = h(A^*_{G^*})$ for $A^* \in \mathcal{B}$ with $\mathcal{B}$

maximal image then choose $\tilde{B}^*$ a basic set such that $h(\tilde{B}^*_{G^*})$ is maximal among the images containing F. By (3) there exists B^* in $\mathcal{B}$ such that $h(\tilde{B}^*_{G^*}) = h(B^*_{G^*})$. Hence, $F = h(A^*_{G^*}) \subset h(\tilde{B}^*_{G^*}) = h(B^*_{G^*})$. Since A^* is assumed to have $\mathcal{B}$ maximal image we have $h(B^*_{G^*}) = h(A^*_{G^*})$ and so $F = h(\tilde{B}^*_{G^*})$, i.e. $F \in \mathcal{F}_g$.

□

Now we introduce a condition which allows us to describe sets on which the restriction of h or h^+ is injective.

Assume $\{z_0^*, \dots, z_m^*\}$ is a G^* chain with, as usual, $z_{i+1} = g(z_i^*)$ for $i = 0, \dots, m$ and $z_0 = j(z_0^*)$. If $\dim z_0 = \dim z_{m+1}$ then by Proposition 4a, z_i^* is an expansion simplex with $z_i = j(z_i^*)$ for $i = 0, \dots, m$. Such a chain is called an *interior chain* if it satisfies the following *interior condition*:

$$(4.33) \qquad \cap_{i=0}^{m} g^{-i}(z_i^*) \subset (z_0)^\circ.$$

Although we usually demand that a chain have length at least one, we regard a single element $z^* \in L^*$ as an interior chain if z^* is an expansion simplex with $z^* \subset (z)^\circ$ where $z = j(z^*)$.

9 Lemma. *Let $\{z_0^*, \dots, z_m^*\}$ be a G^* chain with $\dim z_0 = \dim z_{m+1}$ where $z_0 = j(z_0^*)$ and $z_{m+1} = g(z_m^*)$.*

(a) If for some i_0, i_1 such that $0 \le i_0 \le i_1 \le m$ the chain $\{z_{i_0}^*, \dots, z_{i_1}^*\}$ is an interior chain then $\{z_0^*, \dots, z_m^*\}$ is an interior chain.

(b) Let $\mathbf{z}^* \in K_{G^*}^{*+}$ with $z_0 = j(\mathbf{z}_0^*)$. Assume that $\dim z_0 = \dim \mathbf{z}_i^*$ for all $i \in \mathbf{Z}_+$ and $x = h^+(\mathbf{z}^*)$ is the unique point of X such that $g^i(x) \in \mathbf{z}_i^*$ for all $i \in \mathbf{Z}_+$. There exists m such that $\{\mathbf{z}_0^*, \dots, \mathbf{z}_m^*\}$ is an interior chain iff $x \in (z_0)^\circ$.

Proof. (a) Let $m_0 = i_1 - i_0$ and $m_1 = m - i_0$. If $\cap_{i=0}^{m_0} g^{-i}(z_{i_0+i}^*) \subset (z_{i_0})^\circ$ then shrinking further to $\cap_{i=0}^{m_1} g^{-i}(z_{i_0+i}^*)$ we remain in $(z)_{i_0}^\circ$. Now pull back by the linear bijections $g : z_0^* \to z_1, \dots, z_{i_0-1}^* \to z_{i_0}$. We obtain the left side of (4.33) contained in $(z_0)^\circ$.

(b) $\{C_m = \cap_{i=0}^{m} g^{-i}(\mathbf{z}_i^*) : m \in \mathbf{Z}_+\}$ is a decreasing sequence of compacta in z_0 with intersection x. $\{\mathbf{z}_0^*, \dots, \mathbf{z}_m^*\}$ is an interior chain exactly when $C_m \subset (z_0)^\circ$ and this certainly implies $x \in (z_0)^\circ$. On the other hand, $(z_0)^\circ$ is open in z_0 and so if the intersection x is in $(z_0)^\circ$ then for sufficiently large m $C_m \subset (z_0)^\circ$. □

For $\mathbf{z}^* \in K^{*+}_{G^*}$ with $z_0 = j(z^*_0)$ we call $\mathbf{z}^*$ an *interior element* if for every $i \in \mathbf{Z}_+$, $\dim z_0 = \dim z^*_i$ and there exists $n_i \geq 0$ such that $\{z^*_i, \ldots, z^*_{i+n_i}\}$ is an interior chain. Similarly, $\mathbf{z}^* \in K^*_{G^*}$ is called an *interior element* if dim z^*_i is independent of $i \in \mathbf{Z}$ and for every $i \in \mathbf{Z}$ there exists $n_i \geq 0$ such that $\{z^*_i, \ldots, z^*_{i+n_i}\}$ is an interior chain. The constant dimension conditions imply that the z^*_i's are all expansion simplices and so are nondegenerate. When the constant dimension is k we call $\mathbf{z}^*$ a *dimension k interior element.*

Let Int^+_k and Int_k denote the sets of dimension k interior elements in $K^{*+}_{G^*}$ and $K^*_{G^*}$, respectively. From (4.14):

$$\mathrm{Int}^+_k \subset S^k(K^*)^+_{G^*} \subset ND(K^*)^+_{G^*}$$

and

$$\mathrm{Int}_k \subset S^k(K^*)_{G^*} \subset ND(K^*)_{G^*}. \tag{4.34}$$

In particular, if $\mathbf{z}^* \in \mathrm{Int}^+_k$ then $h^+(\mathbf{z}^*)$ is the unique point x of X such that $g^i(x) \in z^*_i$ for all $i \in \mathbf{Z}_+$. Similarly, if $\mathbf{z}^* \in \mathrm{Int}_k$ then $h(\mathbf{z}^*)$ is the unique point ξ of X_g such that $\xi_i \in z^*_i$ for all $i \in \mathbf{Z}$.

10 Proposition. *(a) Assume $x \in X$ and $\mathbf{z}^* \in K^{*+}_{G^*}$ with dim $z^*_0 \leq k$. The following conditions are equivalent:*

(1) $\mathbf{z}^* \in \mathrm{Int}^+_k$ and $x = h^+(\mathbf{z}^*)$.

(2) $g^i(x) \in z^*_i$ and $g^i(x) \in |\overline{S}^k(K)|\backslash|\overline{S}^{k-1}(K)|$ for all $i \in \mathbf{Z}_+$.

(3) dim $j(z^*_0) = k =$ dim z^*_i for all $i \in \mathbf{Z}_+$ and $\mathbf{z}^* = q^{*+}(x)$.

(b) Assume $\xi \in X_g$ and $\mathbf{z}^* \in K^*_{G^*}$ with dim $z^*_i = k$ for all $i \in \mathbf{Z}$. The following conditions are equivalent:

(1) $\pi^+_{G^*}(\mathbf{z}^*) \in \mathrm{Int}^+_k$ and $\xi_i \in z^*_i$ for all $i \in \mathbf{Z}$.

(2) $\mathbf{z}^* \in \mathrm{Int}_k$ and $\xi = h(\mathbf{z}^*)$.

(3) $\xi_i \in z^*_i$ and $\xi_i \notin |\overline{S}^{k-1}(K)|$ for all $i \in \mathbf{Z}$.

(4) $\mathbf{z}^* = q^*(\xi)$.

Proof. (a) (1) $\Rightarrow$ (2): If $\mathbf{z}^* \in \mathrm{Int}^+_k$ then $k = \dim j(z^*_0)$ and by Proposition 4a $g(z^*_i) = j(z^*_{i+1})$ has dimension k for all $i \in \mathbf{Z}_+$. Hence, for all $i \in \mathbf{Z}_+$, $(s^+_{G^*})^i(\mathbf{z}^*) \in \mathrm{Int}^k_+$ and so $g^i(x) = h^+((s^+_{G^*})^i(\mathbf{z}^*)) \in |\overline{S}^k(K)|\backslash|\overline{S}^{k-1}(K)|$ by

Lemma 9b, when $x = h^+(\mathbf{z}^*)$. That $g^i(x) \in z_i^*$ for all i follows from the definition of the map h^+.

(2) $\Rightarrow$ (3): $k \geq \dim z_0^* \geq \dim z_i^*$ for all $i \in \mathbf{Z}_+$. If $g^i(x) \in |\overline{S}^{k-1}(K^*)|$ for some i then $g^{i+1}(x) \in |\overline{S}^{k-1}(K)|$. So by assumption (2), $g^i(x) \in z_i^* \backslash |\overline{S}^{k-1}(K^*)|$. Hence, dim $z_i^* \geq k$, implying dim $z_i^* = k$. So $g^i(x) \notin \partial z_i^*$. Thus, $q^*(g^i(x)) = z_i^*$. In particular, $j(z_0^*) = j(q^*(x)) = q(x)$ has dimension at least k. But since $x \in |\overline{S}^k(K)|$, dim $q(x) \leq k$.

(3) $\Rightarrow$ (1): Apply Lemma 9b to get $\mathbf{z}^* \in \mathrm{Int}_k^+$.

(b) (1) $\Rightarrow$ (2): By Lemma 9a, $\{z_0^*, \ldots, z_m^*\}$ interior implies $\{z_{-i}^*, \ldots, z_m^*\}$ for all $i \in \mathbf{Z}_+$.

(2) $\Rightarrow$ (3) and (3) $\Rightarrow$ (4): Just like (a). Or apply (a) to $\pi_{G^*}^+((s_{G^*})^i(\mathbf{z}^*))$ for each $i \in \mathbf{Z}$.

(4) $\Rightarrow$ (1): Apply (3) $\Rightarrow$ (1) of (a). □

11 Corollary *For* $0 \leq k \leq \dim, X$.

(a) Int_k^+ *is a + invariant* G_δ *subset of* $K_{G^*}^{*+}$ *contained in the closed invariant set* $S^k(K^*)_{G^*}^+$. *In fact, using the restriction of* h^+ *to* $S^k(K^*)_{G^*}^+$ *we have*

$$(4.35) \qquad \mathrm{Int}_k^+ = (h^+|S^k(K^*)_{G^*}^+)^{-1}(\cap_{i=0}^{\infty} g^{-i}(|\overline{S}^k(K)|\backslash|\overline{S}^{k-1}(K)|)).$$

Thus, $h^+|S^k(K^*)_{G^*}^+$ *is injective over the image* $h^+(\mathrm{Int}_k^+)$ *and* $h^+|\mathrm{Int}_k^+$ *is a homeomorphism onto its image* $\cap_{i=0}^{\infty} g^{-i}(|\overline{S}^k(K)|\backslash|\overline{S}^{k-1}(K)|)$ *with inverse* q^{*+}.

(b) Int_k *is an invariant* G_δ *subset of* $K_{G^*}^*$ *contained in the closed invariant set* $S^k(K^*)_{G^*}$. *Using the restriction of* h *to* $S^k(K^*)_{G^*}$ *we have*

$$(4.36)\ \mathrm{Int}_k = (h|S^k(K^*)_{G^*})^{-1}\{\xi \in X_g : \xi_i \in X\backslash|\overline{S}^{k-1}(K)| \text{ for all } i \in \mathbf{Z}\}.$$

Thus, $h|S^k(K^*)_{G^*}$ *is injective over the image* $h(\mathrm{Int}_k)$ *and* $h|\mathrm{Int}_k$ *is a homeomorphism onto its image* $\{\xi \in X_g : \xi_i \in |\overline{S}^k(K)|\backslash|\overline{S}^{k-1}(K)| \text{ for all } i \in \mathbf{Z}\}$ *with inverse* q^*.

Proof. (a) If $\mathbf{z}^* \in \mathrm{Int}_k^+$ then the proof of (1) $\Rightarrow$ (2) in Proposition 10a includes the result that $(s_{G^*}^+)^i(\mathbf{z}^*) \in \mathrm{Int}_k^+$ for all $i \in \mathbf{Z}_+$. Also, with $x = h^+(\mathbf{z}^*)$ that implication says that $x \in \cap_{i=0}^{\infty} g^{-i}(|\overline{S}^k(K)|\backslash|\overline{S}^{k-1}(K)|)$. Conversely, if x is in this intersection then $q(x) = j(q^*(x))$ has dimension k and so dim $q^*(x) \leq k$. By (2) $\Rightarrow$ (3) and (2) $\Rightarrow$ (1) of Proposition 10a,

$q^{*+}(x) \in \mathrm{Int}_k^+$ and it is the unique $\mathbf{z}^* \in \overline{S}^k(K^*)^+_{G^*}$ such that $g^i(x) \in z_i^*$ for all $i \in \mathbf{Z}_+$. Equation (4.35) then follows and implies that Int_k^+ is a G_δ subset of $S^k(K^*)^+_{G^*}$ and so of $K^{*+}_{G^*}$. Because a continuous map h on a compact space is open at any points z such that $h^{-1}h(z) = \{z\}$, it follows that $h^+|\mathrm{Int}_k^+$ is a homeomorphism onto its image. □

As we saw in Propositions 7 and 8 the periodic elements are of special importance. If $\{z_0^*, \ldots, z_m^*\}$ is a G^* chain with $z_0^* = z_m^*$ then we define the *periodic extension* $\mathbf{z}^*$ in $K^*_{G^*}$ of this chain by:

$$\mathbf{z}_i^* = z_i^* \quad i = 0, \ldots, m$$

$$\mathbf{z}_{i+m}^* = \mathbf{z}_i^* \quad \text{for all } i \in \mathbf{Z}. \tag{4.37}$$

We call the projection $\pi^+_{G^*}(\mathbf{z}^*)$ in $K^{*+}_{G^*}$ the *periodic extension* in $K^{*+}_{G^*}$. By Proposition 4b the z_i^*'s are all expansion simplices of the same dimension.

Similarly, if $x \in X$ satisfies $g^m(x) = x$ we define the *periodic orbit sequence* ξ in X_g by

$$\begin{aligned} \xi_i &= g^i(x) \quad \text{for all } i \in \mathbf{Z}_+ \\ \xi_{i+m} &= \xi_i \quad \text{for all } i \in \mathbf{Z}. \end{aligned} \tag{4.38}$$

12 Corollary. *(a) Let $\{z_0^*, \ldots, z_m^*\}$ be a periodic G^* chain of dimension k, i.e. $z_0^* = z_m^*$ with $k = \dim z_0^*$. Let $\mathbf{z}^*$ be the periodic extension to an element of $K^*_{G^*}$ and $\pi^+_{G^*}(\mathbf{z}^*)$ the extension to $K^{*+}_{G^*}$. If $x = h^+(\pi^+_{G^*}(\mathbf{z}^*))$ then x is the unique point of X such that $g^i(x) \in z_i^*$ for $i = 0, \ldots, m$ and $g^m(x) = x$. Furthermore, the following conditions are equivalent:*

(1) $\mathbf{z}^* \in \mathrm{Int}_k$.

(2) $\pi^+_{G^*}(\mathbf{z}^*) \in \mathrm{Int}_k^+$.

(3) $\pi^+_{G^*}(\mathbf{z}^*) = q^{*+}(x)$.

(4) $g^i(x) \in (z_i)^\circ$ for $i = 0, \ldots, m$.

(5) $x \in X \setminus |\overline{S}^{k-1}|$.

(b) Let x be a point of $|\overline{S}^k(K)| \setminus |\overline{S}^{k-1}(K)|$ such that $g^m(x) = x$. Let ξ be the periodic orbit sequence of x in X_g. If for $i = 0, 1, \ldots, m$, $z_i^* = q^*(g^i(x))$

then $\{z_0^*, \ldots, z_m^*\}$ *is a periodic* G^* *chain of dimension* k. *Its periodic extension is* $q^*(\xi)$ *a dimension* k *interior element of* $K_{G^*}^*$, *with* $\pi_{G^*}^+(q^*(\xi)) = q^{*+}(x)$ *a dimension* k *interior element of* $K_{G^*}^{*+}$.

Proof. (a) Because $g^m(x) = h^+((s_{G^*}^+)^m(\pi_{G^*}^+(\mathbf{z}^*))) = h^+(\pi_{G^*}^+(\mathbf{z}^*)) = x$ it follows that x satisfies the required conditions. On the other hand, if $g^m(\tilde{x}) = \tilde{x}$ and $g^i(\tilde{x}) \in z_i^*$ for $i = 0, \ldots, m$ then $g^i(\tilde{x}) \in \mathbf{z}_i^*$ for all $i \in \mathbf{Z}_+$ and so $\tilde{x} = h^+(\pi_{G^*}^+(\mathbf{z}^*)) = x$. Thus, x is uniquely characterized by these conditions.

(1) $\Rightarrow$ (2): By (2) $\Rightarrow$ (1) of Proposition 10b.

(2) $\Rightarrow$ (3): By (1) $\Rightarrow$ (3) of Proposition 10a.

(3) $\Rightarrow$ (4) and (4) $\Rightarrow$ (5): Obvious.

(5) $\Rightarrow$ (1): Let ξ be the periodic orbit sequence for x. Clearly, $\xi_i \in \mathbf{z}_i^*$ for all $i \in \mathbf{Z}$. If $\xi_i \in |\overline{S}^{k-1}(K)|$ for some i then $\xi_j \in |\overline{S}^{k-1}(K)|$ for all $j \geq i$. But $\xi_j = x$ for arbitrarily large j and $x \notin |\overline{S}^{k-1}(K)|$. Hence, $\xi_i \in X \backslash |\overline{S}^{k-1}(K)|$ for all i. Apply (3) $\Rightarrow$ (2) of Proposition 10b.

(b) By Lemma 3.1, $\{z_0^*, \ldots, z_m^*\}$ is a G^* chain, $z_m^* = q^*(g^m(x)) = q^*(x) = z_0^*$. As $x \in (z_i^*)^\circ \cap |\overline{S}^k(K)| \backslash |\overline{S}^{k-1}(K)|$ we have dim $z_0^* = k$. If $\mathbf{z}^*$ is the periodic extension then $\xi_i \in (z_i^*)^\circ$ for all $i \in \mathbf{Z}$ by periodicity. Thus, $\mathbf{z}^* = q^*(\xi)$ and $\mathbf{z}^* \in \mathrm{Int}_k$ by part (a). □

13 Proposition. *let* B^* *be a* k *skeleton basic set for* G^* *and let* B *be the associated* k *skeleton basic set for* G.

(a) The following conditions are equivalent and when they hold we call B^* *an* interior *basic set for* G^*.

(1) $h^+(B_{G^*}^{*+}) \cap (X \backslash |\overline{S}^{k-1}(K)|) \neq \emptyset$.

(2) $B_{G^*}^{*+} \cap \mathrm{Int}_k^+ \neq \emptyset$.

(3) $B_{G^*}^* \cap \mathrm{Int}_k \neq \emptyset$.

(4) There exists $\{z_0^*, \ldots, z_m^*\}$ *a* G^* *chain in* B^* *which satisfies the interior condition (4.33).*

(5) For some $x \in X$, B^* *is the endset of* x.

(b) The following conditions on a point $x \in X$ *are equivalent:*

(1) $x \in h^+(B_{G^*}^{*+} \cap \mathrm{Int}_k^+)$.

(2) $x \in h^+(B_{G^*}^{*+}) \cap h^+(\mathrm{Int}_k^+)$.

(3) $q^{*+}(x) \in B^{*+}_{G^*} \cap \mathrm{Int}^+_k$.

(4) $g^i(x) \in |B| \backslash |\overline{S}^{k-1}(K)|$ *for all* $i \in \mathbf{Z}_+$.

(5) $q(g^i(x)) \in B$ *for all* $i \in \mathbf{Z}_+$.

(6) B^* *is the endset of* x *and for some* k *skeleton basic set* $\tilde{B}$ *for* G, $g^i(x) \in |\tilde{B}|$ *for all* $i \in \mathbf{Z}_+$.

(7) $x \in |B|$ *and* B^* *is the endset of* x.

(8) B^* *is the endset of* x *and* x *is a nonwandering point for the restricted map* $g||\overline{S}^k(K^*)|$.

(c) The following conditions on a point $\xi \in X_g$ *are equivalent:*

(1) $\xi \in h(B^*_{G^*} \cap \mathrm{Int}_k)$.

(2) $\xi \in h(B^*_{G^*}) \cap h(\mathrm{Int}_k)$.

(3) $q^*(\xi) \in B^*_{G^*} \cap \mathrm{Int}_k$.

(4) $\xi_i \in |B| \backslash |\overline{S}^{k-1}(K)|$ *for all* $i \in \mathbf{Z}$.

(5) $q(\xi_i) \in B$ *for all* $i \in \mathbf{Z}$.

(6) B^* *is the endset of* ξ *and for some* k *skeleton basic set* $\tilde{B}$ *for* G $\xi_i \in |\tilde{B}|$ *for all* $i \in \mathbf{Z}$.

(7) B^* *is both the endset of* ξ *and the startset of* ξ.

(8) B^* *is the endset of* ξ *and* ξ *is a nonwandering point for the restricted homeomorphism* $s_g|(X_g \cap |\overline{S}^k(K^*)|^{\mathbf{Z}})$.

Proof. (a) (1) $\Rightarrow$ (4): If $\mathbf{z}^* \in B^{*+}_{G^*}$ and $x = h^+(\mathbf{z}^*) \notin |\overline{S}^{k-1}(K)|$, then with $z_0 = j(z_0^*)$ $x \in z_0^*$ implies $x \in z_0$. Since $z_0^* \in B^*$ it is an expansion simplex and so z_0 has dimension k. As $x \notin z_0 \cap |\overline{S}^{k-1}(K)| = \partial z_0$, $x \in (z_0)^\circ$. So by Lemma 9b, $\{z_0^*, \ldots, z_m^*\}$ is an interior chain for m large enough.

(4) $\Rightarrow$ (3): By extending the G^* chain in the $\mathcal{O}G^* \cap \mathcal{O}G^{*-1}$ equivalence class B^* and then applying Lemma 9a, we can assume that $z_m^* = z_0^*$. Let $\mathbf{z}^*$ be the periodic extension to an element of $B^*_{G^*}$. By periodicity and Lemma 9a, again, $\mathbf{z}^* \in \mathrm{Int}_k$.

(3) $\Rightarrow$ (2): If $\mathbf{z}^* \in B^*_{G^*} \cap \mathrm{Int}_k$ then $\pi^+_{G^*}(\mathbf{z}^*) \in B^{*+}_{G^*} \cap \mathrm{Int}^+_k$.

(2) $\Rightarrow$ (5): If $\mathbf{z}^* \in B^{*+}_{G^*} \cap \mathrm{Int}^+_k$ and $x = h(\mathbf{z}^*)$ then by Proposition 10a ((1)$\Rightarrow$(3)) $\mathbf{z}^* = q^{*+}(x)$. So B^* is the endset of x.

(5) $\Rightarrow$ (1): Suppose $q^*(g^i(x)) \in B^*$ for $i \geq n$. Then $q^{*+}(g^n(x)) \in B^{*+}_{G^*}$ and so $g^n(x) \in h^+(B^{*+}_{G^*})$. Because $z^*_n \equiv q^*(g^n(x)) \in B^*$ dim $z^*_n = k$ and $g^n(x) \in (z^*_n)^\circ \subset X \backslash |\overline{S}^{k-1}(K)|$.

(b) (1) $\Rightarrow$ (2): Obvious.

(2) $\Rightarrow$ (3): If $x = h^+(\mathbf{z}^*)$ and $\mathbf{z}^* \in \mathrm{Int}^+_k$ then by Proposition 10a ((1) $\Rightarrow$ (3)) $q^{*+}(x) \in \mathrm{Int}^+_k$. If $x = h^+(\mathbf{z}^*)$ with $\mathbf{z}^* \in B^{*+}_{G^*}$ and $q^{*+}(x) \in \mathrm{Int}^+_k$ then $x \in h^+(\mathrm{Int}^+_k)$ and $\mathbf{z}^* \in \overline{S}^k(K^*)^+_{G^*}$ imply $\mathbf{z}^* = q^{*+}(x)$ by Corollary 11a. Thus, $q^{*+}(x) \in B^{*+}_{G^*} \cap \mathrm{Int}^+_k$.

(3) $\Rightarrow$ (4): By Proposition 10a ((3) $\Rightarrow$ (2)), dim $z_0 = k = \dim z^*_i$ with $\mathbf{z}^* = q^{*+}(x)$ and $z_0 = j(z^*_0)$ implies $g^i(x) \in X\backslash|\overline{S}^{k-1}(K)|$ for all $i \in \mathbf{Z}_+$. Clearly, $g^i(x) \in q^*(g^i(x)) \in B^*$ implies $g^i(x) \in |B|$ for all $i \in \mathbf{Z}_+$.

(4) $\Rightarrow$ (5): Let $z_i \in B$ with $g^i(x) \in z_i$. Since dim $z_i = k$, (4) implies $g^i(x) \notin \partial z_i$ and so $g^i(x) \in (z_i)^\circ$, i.e. $q(g^i(x)) = z_i$ for all $i \in \mathbf{Z}_+$.

(5) $\Rightarrow$ (6): Since $q(g^i(x)) = j(q^*(g^i(x)))$ has dimension k for all $i \in \mathbf{Z}_+$ the G^* chain $q^{*+}(x)$ consists of expansion simplices by Proposition 4a. Hence, with $z^*_i = q^*(g^i(x))$ we have that $j(z^*_i)$ and $g(z^*_i) = j(z^*_{i+1}) \in B$. So by (4.10), $z^*_i \in B^*$ for all i. Hence, B^* is the endset of x. Clearly, for all $i \in \mathbf{Z}_+$ $g^i(x) \in |\tilde{B}|$ with $\tilde{B} = B$.

(6) $\Rightarrow$ (7): For $i \in \mathbf{Z}_+$, let $z_i \in \tilde{B}$ with $g^i(x) \in z_i$, so that $q(g^i(x)) < z_i$. As B^* is the endset of x, there exists i_0 so that $q^*(g^i(x)) \in B^*$ for $i \geq i_0$. Hence, for $i \geq i_0$, $q(g^i(x)) = j(q^*(g^i(x))) \in B$ by (4.9). Because $k = \dim z_i = \dim q(g^i(x))$ it follows that $z_i = q(g^i(x))$ for $i \geq i_0$. Because distinct basic sets for G are disjoint, $B = \tilde{B}$. In particular, $x \in z_0 \in \tilde{B} = B$ and so $x \in |B|$.

(7) $\Rightarrow$ (1): Let $\mathbf{z}^* = q^{*+}(x)$ and let i_0 be such that $z^*_i = q^*(g^i(x)) \in B^*$ for $i \geq i_0$. Let $z_i = j(q^*(g^i(x)))$, so that $z_i \in B$ for $i \geq i_0$. Since B^* and B consist of k simplices, dim $z_i = k$ and dim $z^*_i = k$ for $i \geq i_0$ and so by (4.6) dim $z_0 \geq \dim z^*_0 \geq \dim z^*_i \geq k$ for all $i \in \mathbf{Z}_+$. Since $x \in |B|$, there exists $w \in B$ such that $z_0 = q(x) < w$. But then $k \geq \dim w \geq \dim z_0 \geq k$ implies $w = z_0$ and so $z_0 \in B$ and dim $z_i = \dim z^*_i = k$ for all i. Since $\mathbf{z}$ is a G chain with $z_0 \in B$ and $z_i \in B$ for all $i \geq i_0$, it follows that for all $i \in \mathbf{Z}_+$, z_i is in the $\mathcal{O}G \cap \mathcal{O}(G^{-1}$ equivalence class B. Because $j(z^*_i) = z_i \in B$ and, by Proposition 4a, $g(z^*_i) = j(z^*_{i+1}) = z_{i+1} \in B$ it follows from (4.10) that $\mathbf{z}^* = q^{*+}(x) \in B^{*+}_{G^*}$. By Proposition 10a ((3) $\Rightarrow$ (1)) $q^{*+}(x) \in \mathrm{Int}^+_k$. Hence, $x = h^+(q^{*+}(x)) \in h^*(B^{*+}_{G^*} \cap \mathrm{Int}^+_k)$.

This completes the proof that (1) - (7) are equivalent.

(1) $\Rightarrow$ (8): The subshift map $s^+_{G^*}$ on the basic set $B^{*+}_{G^*}$ is topologically transitive and so g is topologically transitive on $h^+(B^{*+}_{G^*})$. In particular, every point x of $h^+(B^{*+}_{G^*})$ is nonwandering for the restriction of g to $h^+(B^{*+}_{G^*})$ and a fortiori for the restriction to $|\overline{S}^k(K^*)|$. B^* is the endset of x because (1) $\Rightarrow$ (7).

(8) $\Rightarrow$ (7): As in the proof of (7) $\Rightarrow$ (1), let $\mathbf{z}^* = q^{*+}(x)$ and let i_0 be such that $z^*_i = q^*(g^i(x))$ is in the endset B^* for all $i \geq i_0$. As $x \in |\overline{S}^k(K^*)|$, $z^*_0 = q^*(x)$ has dimension at most k. Hence, by (4.6) $k \geq \dim z^*_0 \geq \dim z^*_i \geq k$ for all $i \in \mathbf{Z}_+$. So for all $i \in \mathbf{Z}_+$ $\dim z^*_i = k$. Since $g^i(x) \in (z^*_i)^\circ$ it follows that $(z^*_i)^\circ$ is a neighborhood of $g^i(x)$ in $|\overline{S}^k(K^*)|$ for all $i \in \mathbf{Z}_+$. Define

$$U = \cap_{i=0}^{i_0} g^{-i}((z^*_i)^\circ). \tag{4.39}$$

Thus U is a neighborhood of x in $|\overline{S}^k(K^*)|$. Because x is nonwandering for the restriction of g to $|\overline{S}^k(K^*)|$ there exists $\tilde{x}$ and $i_1 > i_0$ such that $\tilde{x}$, $g^{i_1}(\tilde{x}) \in U$. Let $\tilde{\mathbf{z}}^* = q^{*+}(\tilde{x})$. So $\{\tilde{z}^*_0, \dots, \tilde{z}^*_{i_1}\}$ is a G^* chain. Because $\tilde{x}$, $g^{i_1}(\tilde{x})$ are in the subset U of $(z^*_0)^\circ$ it follows that

$$\tilde{z}^*_0 = \tilde{z}^*_{i_1} = z^*_0 \tag{4.40}$$

and so the chain $\{\tilde{z}^*_0, \dots, \tilde{z}^*_{i_1}\}$ is periodic. It thus consists of simplices in some G^* basic set $\tilde{B}^*$. But by definition of U

$$\tilde{z}^*_i = q^*(g^i(\tilde{x})) = z^*_i \quad \text{for } i \leq i_0. \tag{4.41}$$

In particular, $\tilde{z}^*_{i_0} = z^*_{i_0} \in B^*$. Since distinct basic sets are disjoint $B^* = \tilde{B}^*$. It follows that $z^*_0 = \tilde{z}^*_0 \in B^*$ and so $x \in |B^*|$ which is a subset of $|B|$, completing the proof of (7).

(c) The proofs of (1) $\Rightarrow$ (2) $\Rightarrow$ (3) $\Rightarrow$ (4) $\Rightarrow$ (5) $\Rightarrow$ (6) are completely analogous to the corresponding proofs in part (b), using Proposition 10b and Corollary 11b in place of Proposition 10a and Corollary 10a.

(6) $\Rightarrow$ (7): Assumption (6) implies ξ_i satisfies (6) of part (b) and so by (3) of (b), $q^*(\xi_i) \in B^*$ for all i. Hence, B^* is the startset for ξ.

(7) $\Rightarrow$ (1): By assumption (7), the G^* chain $q^*(\xi)$ lies in B^* for $|i|$ large enough. Since B^* is an $\mathcal{O}G^* \cap \mathcal{O}G^{*-1}$ equivalence class the intermediate terms lie in B^* as well. Thus, $q^*(\xi) \in B^*_{G^*}$. By Proposition 10b ((4) $\Rightarrow$ (3) applied to $\mathbf{z}^* = q^*(\xi)$) $q^*(\xi) \in \mathrm{Int}_k$. Hence, $\xi = h(q^*(\xi))$ is in $h(B^*_{G^*} \cap \mathrm{Int}_k)$.

(1) $\Rightarrow$ (8): Just as in part (b).

(8) $\Rightarrow$ (5): $X_g \cap |\overline{S}^k(K^*)|^{\mathbf{Z}}$ is an s_g invariant set and so the restriction of s_g to it is a homeomorphism. Hence, ξ nonwandering for the restriction implies $s_g^i(\xi)$ is nonwandering as well. Projecting by π_0 to X we have that ξ_i is nonwandering for $g||\overline{S}(K^*)|$. Since the endset of ξ_i is the endset of ξ we see that ξ_i satisfies (8) of (b). So by (5) of (b), $q(\xi_i) \in B$ for all i, i.e. (5) holds. □

A closed + invariant subset A of X is a topologically transitive subset, or just a *transitive subset*, if the restriction $g|A : A \to A$ is a topologically transitive map. That is, there exists $x \in A$ such that $\omega g(x) = A$. The subset

(4.42) $$\text{Trans}_{g|A} = \{x \in A : \omega g(x) = A\}$$

is then a dense G_δ subset of A (see, e.g., Akin (1993) Theorem 4.12). Clearly, $x \in \text{Trans}_{g|A}$ implies $x \in \omega g(x)$ and so x is a recurrent point for g. Conversely, if x is a recurrent point, $x \in \omega g(x)$, then $A = \omega g(x)$ is a transitive subset with $x \in \text{Trans}_{g|A}$. In general, if x is a recurrent point and A is a closed + invariant subset of X such that $x \in A$ then $x \in \omega g(x) \subset A$ implies x is a nonwandering point for $g|A$.

14 Corollary. *Let A be a transitive subset for the p.l. map g on X. There exists a basic set B^* for G^* such that:*

$$A \subset h^+(B^{*+}_{G^*}),$$

(4.43) $$\text{Trans}_{g|A} \subset h^+(B^{*+}_{G^*} \cap \text{Int}^+_k),$$

where $k = \dim z^*$ for all z^* in B^*. For all $x \in \text{Trans}_{g|A}$, the interior basic set B^* is the endset of x. If $\tilde{B}^*$ is a basic set for G^* and there exists $\mathbf{z}^*$ an interior chain in $\tilde{B}^{*+}_{G^*}$ such that $h^+(\mathbf{z}^*) \in \text{Trans}_{g|A}$ then $\tilde{B}^* = B^*$.

Proof. Fix $x \in \text{Trans}_{g|A}$, let B^* be the endset of x so that B^* is a k skeleton basic set where $k = \dim z^*$ for $z^* \in B^*$. By Proposition 6a, $A = \omega g(x) \subset h^*(B^{*+}_{G^*}) \subset |\overline{S}^k(K)|$. By Proposition 13b ((8) $\Rightarrow$ (4)), $x \in A \backslash |\overline{S}^{k-1}(K)|$. If $g^i(x_1) \in |\overline{S}^{k-1}(K)|$ for some $i \in \mathbf{Z}_+$ then + invariance of $|\overline{S}^{k-1}(K)|$ implies $\omega g(x_1) \subset |\overline{S}^{k-1}(K)|$. Contrapositively, if x_1 is any transitive point for $g|A$ then $g^i(x_1) \in A\backslash|\overline{S}^{k-1}(K)|$ for all $i \in \mathbf{Z}_+$. By Proposition 13b ((4) $\Rightarrow$ (7), (11)) B^* is the endset of x_1 and $x_1 \in h^+(B^{*+}_{G^*} \cap$

Int_k^+). This proves the second inclusion in (4.43). We also have that B^* is the endset of every transitive point.

If $\mathbf{z}^*$ is an interior chain in $K_{G^*}^{*+}$ and $\tilde{x} = h^+(\mathbf{z}^*)$ then by Proposition 10a ((1) $\Rightarrow$ (3)), $\mathbf{z}^* = q^{*+}(\tilde{x})$. If $\mathbf{z}^* \in \tilde{B}_{G^*}^{*+}$ as well, then $q^*(g^i(\tilde{x})) \in \tilde{B}^*$ for all $i \in \mathbf{Z}_+$ implies $\tilde{B}^*$ is the endset of $\tilde{x}$. Then $\tilde{x} \in \mathrm{Trans}_{g|A}$ implies $\tilde{B}^* = B^*$.

□

Recurrence of a point x in X is an intrinsic property. That is, if A is a closed + invariant subset of X and $x \in A$ then x is recurrent for g, i.e. $x \in \omega g(x)$ iff x is recurrent for the restriction $g|A$. If x is nonwandering for $g|A$ then x is nonwandering for g. Proposition 13b implies that if $q^{*+}(x)$ is a G^* chain of constant dimension k, and x is nonwandering for the restriction $g||\overline{S}^k(K^*)|$ then x lies in some basic set image. In fact, $x \in h^+(B_{G^*}^{*+} \cap \mathrm{Int}_k^+)$ where B^* is the endset of x. We will later see examples where $q^{*+}(x)$ consists of k simplices and x is nonwandering for g, but x does not lie in any basic set image.

15 Theorem. *Let $g : K^* \to K$ be a simplicial dynamical system with $X = |K^*| = |K|$. Let s_g denote the sample path homeomorphism associated with the p.l. map g on X and let s_{G^*} and $s_{G^*}^+$ denote the finite type sample path shifts associated with the relation G^* on K^*. Let $\pi_{G^*}^+ : K_{G^*}^* \to K_{G^*}^{*+}$ and $\pi_0 : X_g \to X$ be the projection maps. For $\mathbf{z}^* \in K_{G^*}^{*+}$ consisting of nondegenerate simplices, $h^+(\mathbf{z}^*)$ is the unique point x of X such that $g^i(x) \in z_i^*$ for all $i \in \mathbf{Z}_+$. For $\mathbf{z}^* \in K_{G^*}^*$ consisting of nondegenerate simplices, $h(\mathbf{z}^*)$ is the unique point ξ of X_g such that $\xi_i \in z_i^*$ for all $i \in \mathbf{Z}$, so that $\pi_0(h(\mathbf{z}^*)) = h^+(\pi_{G^*}^+(\mathbf{z}^*))$.*

(a) Let B^* be a k skeleton basic set for G^*. The restriction of s_{G^*} to the basic set $B_{G^*}^*$ is topologically transitive with dense periodic points. The same is true for $s_{G^*}^+$ on $B_{G^*}^{*+}$ and for s_g and g on the basic set images $h(B_{G^*}^*)$ and $h^+(B_{G^*}^{*+}) = \pi_0(h(B_{G^*}^*))$ respectively. The continuous surjection $h^+|B_{G^*}^{*+} : B_{G^*}^{*+} \to h^+(B_{G^*}^{*+})$ (and $h|B_{G^*}^* : B_{G^*}^* \to h(B_{G^*}^*)$) maps the shift $s_{G^*}^+$ to g (resp. the shift s_{G^*} to s_g).

If B^* is an interior basic set, i.e. $h^+(B_{G^*}^*) \cap (X \backslash |\overline{S}^{k-1}(K)|) \neq \emptyset$ then $h^+|B_{G^*}^{*+}$ and $h|B_{G^*}^*$ are almost homeomorphisms. Specifically, $h^+(B_{G^*}^{*+} \cap \mathrm{Int}_G^+)$ is a dense G_δ subset of $h^+(B_{G^*}^{*+})$ containing the transitive points of $g|h^+(B_{G^*}^{*+})$ and $\mathbf{z}^* \in B_{G^*}^{*+} \cap \mathrm{Int}_k^+$ implies $(h^+|B_{G^*}^{*+})^{-1}(h(\mathbf{z}^*)) = \mathbf{z}^*$, i.e. $h^+|B_{G^*}^{*+}$ is injective over $h^+(B_{G^*}^{*+} \cap \mathrm{Int}_k^+)$. Similarly, $h(B_{G^*}^* \cap \mathrm{Int}_k)$ is a dense G_δ

subset of $h(B^*_{G^*})$ *containing the transitive points of* $s_g|h(B^*_{G^*})$ *and over which* $h|B^*_{G^*}$ *is injective.*

If B^* *is not an interior basic set then there exists* B^*_1 *a* k_1 *skeleton interior basic set with* $k_1 < k$ *such that* $h^+(B^{*+}_{G^*}) \subset h^+(B^{*+}_{1G^*})$ *and* $h(B^*_{G^*}) \subset h(B^*_{1G^*})$.

(b) Let $\mathcal{F}_g$ *be the family of maximal basic set images in* X_g. *Let* $\mathcal{F}^+_g = \{\pi_0(F) : F \in \mathcal{F}_g\}$, *or, equivalently the set of maximal basic set images in* X. *These families satisfy the following properties:*

(1) If $\xi \in X_g$ *then there exist* F_0, $F_1 \in \mathcal{F}_g$ *such that* $\alpha s_g(\xi) \subset F_0$ *and* $\omega s_g(\xi) \subset F_1$, *and so* $\mathcal{F}_g$ *is a fine pre-decomposition for the homeomorphism* s_g *on* X_g. *If* $x \in X$ *then there exists* $F^+ \in \mathcal{F}^+_g$ *and* $i_0 \in \mathbf{Z}_+$ *such that* $g^i(x) \in F^+$ *for* $i \geq i_0$ *and so* $\omega g(x) \subset F^+$.

(2) If A *is a closed, invariant subset of* X_g *on which* s_g *is topologically transitive (resp. of* X *on which* g *is topologically transitive) then there exists* $F \in \mathcal{F}_g$ *such that* $A \subset F$ *(resp.* $F^+ \in \mathcal{F}^+_g$ *such that* $A \subset F^+$*).*

(3) The union $\cup\mathcal{F}_g$ *(or* $\cup\mathcal{F}^+_g$*), i.e. the union of the basic set images in* X_g *(resp.* X*), is the Birkhoff center for* s_g *(resp. for* g*), that is, the closure of the set of recurrent points.*

(4) $\mathcal{F}_g$ *(or* $\mathcal{F}^+_g$*) is the family of maximal interior basic set images in* X_g *(resp. in* X*).*

Proof. (a) As $B^*_{G^*}$ is the basic set for a subshift of finite type, s_{G^*} on $B^*_{G^*}$ is a topologically transitive homeomorphism with dense periodic points. As $\pi^+_{G^*}$ maps $s_{G^*}|B^*_{G^*}$ to $s^+_{G^*}|B^{*+}_{G^*}$, $h|B^*_{G^*}$ maps $s_{G^*}|B^*_{G^*}$ to $s_g|h(B^*_{G^*})$ and $h^+|B^{*+}_{G^*}$ maps $s^+_{G^*}|B^{*+}_{G^*}$ to $g|h^+(B^{*+}_{G^*})$ it follows that on each of the images the associated dynamical system is topologically transitive with dense periodic points. In particular the transitive points are dense.

Now assume that B^* is an interior k skeleton basic set and that A is the basic set image $h^+(B^{*+}_{G^*})$ If $g^i(x) \in |\overline{S}^{k-1}(K)|$ for some i then $\omega g(x) \subset |\overline{S}^{k-1}(K)|$ and so x is not a transitive point for $g|A$ because $A \backslash |\overline{S}^{k-1}(K)| \neq \emptyset$. By Proposition 13b ((4) $\Rightarrow$ (1)) it follows that $h^+(B^{*+}_{G^*} \cap \mathrm{Int}^+_k)$ contains the dense subset $\mathrm{Trans}_{g|A}$ of A. Condition (4) itself shows that $h^+(B^{*+}_{G^*} \cap \mathrm{Int}^+_k)$ is a G_δ subset. By Corollary 11a, $h^+|B^{*+}_{G^*}$ is injective over this subset. Let Trans_{B^*} denote the set of transitive points of $s^+_{G^*}|B^{*+}_{G^*}$. Since h^+ maps

Trans_{B^*} into $\text{Trans}_{g|A}$, we have

$$(4.44) \qquad \text{Trans}_{B^*} \subset B^{*+}_{G^*} \cap \text{Int}^+_k.$$

As Trans_{B^*} is dense in $B^{*+}_{G^*}$ it follows that $h^+|B^{*+}_{G^*}$ is an almost homeomorphism. The results for $h|B^*_{G^*}$, including the analogue of (4.44), are proved similarly using Proposition 13c and Corollary 11b.

If A is a transitive subset of X then by Corollary 14, $A \subset h^+(\tilde{B}^{*+}_{G^*})$ where $\tilde{B}^*$ is the endset of the transitive points of A. This proves (2) of (b) and (4) of (b) as well. If $A = h^+(B^*)$ and B^* is a k skeleton basic set which is not interior, so that $A \subset |\overline{S}^{k-1}(K)|$ then choose $\mathbf{z}^* \in B^{*+}_{G^*}$ such that $x = h^+(\mathbf{z}^*)$ is a transitive point for $g|A$. For all $i \in \mathbf{Z}_+$, $q^*(g^i(x)) < z^*_i$. For sufficiently large i, $q^*(g^i(x))$ is in the endset $\tilde{B}^*$ which is interior, while for all $i \in \mathbf{Z}_+$, $z^*_i \in B^*$ which is not interior. Thus, $B^* \neq \tilde{B}^*$ and so eventually $\tilde{k} = \dim q^*(g^i(x)) < \dim z^*_i = k$ where $\tilde{B}^*$ is a $\tilde{k}$ skeleton basic set.

This completes the proof of (a).

(b) By Proposition 7a, for basic set images A_1 and A_2 in X_g we have $A_1 \subset A_2$ iff $\pi^+_{G^*}(A_1) \subset \pi^+_{G^*}(A_2)$ in X. Hence, $\mathcal{F}^+_g$ is the set of maximal basic set images in X.

We have already proved properties (2) and (4). Property (1) follows from Proposition 6 which implies that $\mathcal{F}_g$ is a pre-decomposition for s_g. It is a fine pre-decomposition because each $F \in \mathcal{F}_g$ is a topologically transitive subset.

The periodic points for g (resp. s_g) are dense in $\cup\mathcal{F}^+_g$ (resp $\cup\mathcal{F}_g$) and so the union is contained in the center. If x is a recurrent point for g, i.e. $x \in \omega g(x)$ then $\omega g(x)$ is a topologically transitive subset and so $\omega g(x) \subset F^+$ for some $F^+ \in \mathcal{F}^+_g$ by property (2) and so $x \in F^+$. Hence, all recurrent points lie in $\cup\mathcal{F}^+_g$ and so the center is contained in the closed set $\cup\mathcal{F}^+_g$. The proof of property (4) for $\cup\mathcal{F}_g$ is completely similar.

Remark. If B^* is a 0 skeleton basic set then by Proposition 4f, it is a single periodic orbit for the map g restricted to the vertex set $V(K)$. Thus, $B^*_{G^*}$ is a single periodic orbit for s_{G^*} and $\pi^+_{G^*} : B^*_{G^*} \to B^{*+}_{G^*}$, $h : B^*_{G^*} \to h(B^*_{G^*})$ and $h^+ : B^{*+}_{G^*} \to h(B^{*+}_{G^*})$ are all bijections. Each of the associated dynamical systems is a single periodic orbit. B^* is clearly interior. □

A continuous map between compact spaces is called *nondegenerate* if the pre-image of every point of the range is zero-dimensional. For a simplicial map this is true exactly when the restriction to each domain simplex is injective. Thus, $g : K^* \to K$ is nondegenerate when every simplex of K^* is nondegenerate, i.e. when $ND(K^*) = K^*$. If g is nondegenerate then for $k = 0, 1, \ldots, \dim X$

$$g^{-1}(S^k(K)) = S^k(K^*), \tag{4.45}$$

which says that z^* in K^* has dimension k iff $z = g(z^*)$ in K has dimension k.

16 Lemma. *Assume g is nondegenerate. If for $z_0^* \in K^*$ and $z_1^* \in G^*(z_0^*)$ there exists $\tilde{z}_0^* \in K^*$ such that dim $\tilde{z}_0^* = k$ and $z_0^* \subset \tilde{z}_0^*$ then there exists $\tilde{z}_1^* \in G^*(\tilde{z}_0^*)$ such that dim $\tilde{z}_1^* = k$ and $z_1^* \subset \tilde{z}_1^*$.*

Proof. Since z_0^* is a face of $\tilde{z}_0^*$, $z_1 = g(z_0^*)$ is a face of $\tilde{z}_1 = g(\tilde{z}_0^*)$ and $\tilde{z}_1$ has dimension k because g is nondegenerate. By definition of G^* $z_1^* \in J^{-1}(z_1) \subset J^{-1}(\tilde{z}_1)$. Because z_1^* is a simplex of the subdivision $J^{-1}(\tilde{z}_1)$ of the k simplex $\tilde{z}_1$, there exists a k simplex $\tilde{z}_1^*$ of $J^{-1}(\tilde{z}_1)$ such that z_1^* is a face of $\tilde{z}_1^*$. □

If the map g is nondegenerate then $\hat{p}$ and $\hat{p}^+$ are homeomorphisms so that h is a function on $K_{G^*}^*$ and h^+ is on $K_{G^*}^{*+}$. By (4.45) $|S^k(K)|$ is then + invariant for every k and so it is reasonable to restrict attention to spaces X which are *everywhere d dimensional*, that is,

$$X = |S^d(K)| = \cup\{z : \dim z = d\}. \tag{4.46}$$

17 Theorem. *Let $g : K^* \to K$ be a nondegenerate simplicial dynamical system with $X = |K^*| = |K|$ everywhere d dimensional, i.e. every point of X is contained in a d simplex. For $\mathbf{z}^* \in K_{G^*}^{*+}$, $h^+(\mathbf{z}^*)$ is the unique point x of X such that $g^i(x) \in z_i^*$ for all $i \in \mathbf{Z}_+$. For $\mathbf{z}^* \in K_{G^*}^*$, $h(\mathbf{z}^*)$ is the unique point ξ of X_g such that $\xi_i \in z_i^*$ for all $i \in \mathbf{Z}$.*

(a) The restricted continuous maps $h^+|S^d(K^*)_{G^*}^+ : S^d(K^*)_{G^*}^+ \to X$ and $h|S^d(K^*)_{G^*} : S^d(K^))_{G^*} \to X_g$ are almost homeomorphisms. Each is a continuous surjection which is injective over a dense G_δ subset.

(b) The set $\mathcal{F}_g$ of maximal basic set images in X_g is the family of maximal d skeleton basic set images in X_g. Similarly, the set $\mathcal{F}_g^+$ of maximal basic set images in X is the family of maximal d skeleton basic set images in X.

Proof. (a) Let $\mathbf{z}^* \in K_{G^*}^{*+}$. Choose $\tilde{z}_0^* \in S^d(K^*)$ such that $z_0^* < \tilde{z}_0^*$. This is possible because X is everywhere d dimensional so that every simplex is a face of some d simplex. Proceed inductively using Lemma 16 to define $\tilde{\mathbf{z}}^* \in S^d(K^*)_{G^*}^+$ with $z_i^* < \tilde{z}_i^*$ for all $i \in \mathbf{Z}$. Clearly, $h^+(\mathbf{z}^*) = h^+(\tilde{\mathbf{z}}^*)$ and so

$$h^+(S^d(K^*)_{G^*}^+) = h^+(K_{G^*}^{*+}) = X. \tag{4.47}$$

It then follows from Lemma 2 that

$$h(S^d(K^*)_{G^*}) = h(K_{G^*}) = X_g. \tag{4.48}$$

Hence, $h^+|S^d(K^*)_{G^*}^+$ and $h|S^d(K^*)_{G^*}$ are continuous surjections.

By (4.35) it follows that

$$h^+(\mathrm{Int}_d^+) = \cap_{i=0}^{\infty} g^{-i}(X \backslash |\overline{S}^{d-1}(K)|). \tag{4.49}$$

Because X is everywhere d dimensional

$$X \backslash |\overline{S}^{d-1}(K)| = \cup\{(z)^\circ : z \in S^d(K)\} \tag{4.50}$$

is open and dense in X. Because g is nondegenerate induction on i shows that $g^{-i}(|\overline{S}^{d-1}(K)|)$ is $d-1$ dimensional and so is nowhere dense in X for all $i \in \mathbf{Z}_+$. So by (4.49) and the Baire Category Theorem $h^+(\mathrm{Int}_d^+)$ is a dense G_δ subset of X.

To show that h^+ and h are almost homeomorphisms on $S^d(K^*)_{G^*}^+$ and $S^d(K^*)_{G^*}$ respectively, it suffices to show that Int_d^+ and Int_d are dense subsets. We can then apply Corollary 11.

So let $\mathbf{z}^* \in S^d(K^*)_{G^*}^+$ and $i_0 \in \mathbf{Z}_+$ be chosen arbitrarily large. Because $(z_{i_0}^*)^\circ$ is open in X and $h^+(\mathrm{Int}_d^+)$ is dense we can choose x in the intersection. Hence, $q^*(x) = z_{i_0}^*$ and $q^{*+}(x) \in \mathrm{Int}_d^+$. Define $\tilde{\mathbf{z}}^*$ by

$$\tilde{z}_i^* = \begin{cases} z_i^* & i \le i_0 \\ q^*(g^{i-i_0}(x)) & i \ge i_0. \end{cases} \tag{4.51}$$

By Lemma 9a $\tilde{\mathbf{z}}^*$ is an interior chain and by choosing i_0 large enough $\tilde{\mathbf{z}}^*$ is arbitrarily close to $\mathbf{z}^*$.

(b) For $x \in X$, let $x = h(\mathbf{z}^*)$ with $\mathbf{z}^* \in S^d(K^*)_{G^*}^+$ and let B^* be the endset for $\mathbf{z}^*$. B^* is a d skeleton basic set but need not be interior since it need not be the endset of x. If $z_i^* \in B^*$ for $i \ge i_0$ then $g^{i_0}(x) \in h^+(B_{G^*}^{*+})$

and so $\omega g(x) \subset h^+(B^{*+}_{G^*})$. Thus, if x is a transitive point for some basic set image A in X then $A \subset h^+(B^{*+}_{G^*})$. It then follows from Proposition 7 that $\mathcal{F}^+_g$ consists of maximal d skeleton images in X and that this implies that $\mathcal{F}_g$ consists of maximal d skeleton images in X_g. □

The following technical result is useful for studying recurrence.

18 Proposition. *Let B^* be a G^* basic set with B the associated G basic set. Assume $y \in |\mathcal{O}G(B)|$, i.e. there is a G chain $\{w_0, \ldots, w_k\}$ with $w_0 \in B$ and $y \in w_k$.*

(a) Assume $x \in X$ is such that for some $\mathbf{z}^* \in K^{*+}_{G^*}$ with B^* the endset of $\mathbf{z}^*$, $g^i(x) \in z^*_i$ for all $i \in \mathbf{Z}_+$. For every neighborhood U of x in X and positive integer N, there exists $\tilde{x} \in U$ and $m > N$ such that $g^m(\tilde{x}) = y$.

(b) Assume $\xi \in X_g$ is such that for some $\mathbf{z}^* \in K^*_{G^*}$ with B^* the endset of $\mathbf{z}^*$, $\xi_i \in z^*_i$ for all $i \in \mathbf{Z}$. For every neighborhood U of ξ in X_g and positive integer N, There exists $\tilde{\xi} \in U$ and $m > N$ such that $\tilde{\xi}_m = y$. Furthermore, $\tilde{\xi}$ can be chosen so that

$$\mathrm{Lim}_{i \to \infty} d_K(\xi_{-i}, \tilde{\xi}_{-i}) = 0. \tag{4.52}$$

Proof. By extending the chain one step left in B if necessary we can assume $\{w_0, \ldots, w_k\}$ is a G chain with $w_0, w_1 \in B$ and $y \in w_k$. By definition of G there exists a G^* chain $\{w^*_0, \ldots, w^*_{k-1}\}$ with $w_i \in J(w^*_i)$ and $w_{i+1} = g(w^*_i)$ for $i = 0, \ldots, k-1$. By (4.6) $\dim w_0 \geq \dim w^*_0 \geq \dim w_1$ and since $w_0, w_1 \in B$, $\dim w_0 = \dim w_1$. So by Lemma 2.6a, $w_0 = j(w^*_0)$. Thus, $j(w^*_0), g(w^*_0) \in B$ and so $w^*_0 \in B^*$ by (4.10). Since $y \in w_k$, $q^*(y) \subset q(y) < w_k$ and so we can define $w^*_k = q^*(y)$ to obtain a G^* chain $\{w^*_0, \ldots, w^*_k\}$ with $w^*_0 \in B^*$ and $w^*_k = q^*(y)$.

(a) For each $i \in \mathbf{Z}_+$ choose $t_i \in T^{z^*_i}$ so that $g^i(x) \in \langle z^*_i, t_i \rangle$. We thus obtain $(\mathbf{z}^*, \mathbf{t}) \in \hat{K}_{\hat{G}}$ such that $\hat{h}(\mathbf{z}^*, \mathbf{t}) = x$. By continuity of $\hat{h}^+$, for every neighborhood U of x there exists a positive integer N_1 such that $(\tilde{\mathbf{z}}^*, \tilde{\mathbf{t}}) \in \hat{K}_{\hat{G}}$ with $(\tilde{z}^*_i, \tilde{t}^*_i) = (z^*_i, t_i)$ for all $i \leq N_1$ implies $\hat{h}^+(\tilde{z}^*, \tilde{t}) \in U$.

Let N_2 be a positive integer so that $z^*_i \in B^*$ for all $i \geq N_2$.

Now given a positive integer N, let $N_3 = \max(N, N_1, N_2)$. Define $(\tilde{z}^*_i, \tilde{t}_i) = (z^*_i, t_i)$ for $i \leq N_3$. Since $\tilde{z}^*_{N_3} = z^*_{N_3} \in B^*$ we can define $\tilde{z}^*_i \in B^*$ for $N_3 \leq i \leq N_4$ so that $\tilde{z}^*_{N_4} = w^*_0$, the first term of the G^* chain constructed in the first paragraph. Since $\tilde{z}^*_i \in B^*$ for $N_3 \leq i \leq N_4$ there is a unique t_i, such that $(\tilde{z}_i^*, \tilde{t}_i) \in \hat{K}$. Now let $m = N_4 + k$ and define $\tilde{z}_i^* = w_{i-N_4}^*$ for $N_4 \leq i \leq m$. For $N_4 \leq i < m$ choose t_i arbitrarily so that $(\tilde{z}_i^*, t_i) \in \hat{K}$. Finally, since $\tilde{z}_m^* = q^*(y)$ we can complete the chain by defining $(\tilde{z}_i^*, \tilde{t}_i) = \hat{q}(g^{i-m}(y))$ for $i \geq m$.

Let $\tilde{x} = \hat{h}^+(\tilde{\mathbf{z}}^*, \tilde{\mathbf{t}})$. Since $(\tilde{z}_i^*, \tilde{t}_i) = (z_i^*, t_i)$ for $i \leq N_1$, $\tilde{x} \in U$. $g^m(\tilde{x}) = \hat{h}^+(s_{\hat{G}+})^m(\tilde{\mathbf{z}}, \tilde{\mathbf{t}}) = \hat{h}^+\hat{q}^+(y) = y$.

(b) As in (a) choose $(\mathbf{z}^*, \mathbf{t}) \in \hat{K}_{\hat{G}}$ such that $x = \hat{h}(\mathbf{z}^*, \mathbf{t})$ and construct $(\tilde{\mathbf{z}}^*, \tilde{\mathbf{t}}) \in \hat{K}_{\hat{G}}$ so that $(\tilde{z}_i^*, \tilde{t}_i) = (z_i^*, t_i)$ for $i \leq N_3$ and $(\tilde{z}_i^*, \tilde{t}_i) = \hat{q}(g^{i-m}(y))$ for $i \geq m$. Since $(\tilde{z}_i^*, \tilde{t}_i) = (z_i^*, t_i)$ for all negative i, (4.52) follows for $\tilde{\xi} = h(\tilde{\mathbf{z}}^*, \tilde{\mathbf{t}})$. □

19 Corollary. *A point x of X has endset a 0 skeleton basic set iff $g^i(x)$ is a vertex of K for some $i \in \mathbf{Z}_+$. The set of such points is dense in X. Similarly, ξ of X_g has endset a 0 skeleton basic set iff ξ_i is a vertex of K for some $i \in \mathbf{Z}_+$. The set of such points is dense in X_g.*

Proof. If $g^{i_0}(x) \in V(K^*)$ for some i_0 then $g^i(x) \in V(K)$ for all $i > i_0$. So $g^{i_0}(x)$ is a vertex for some i_0 exactly when the endset consists of vertices. For any point $x \in X$ let B^* be the endset of x and let v^* be a vertex of some simplex of B^*. Hence, $v^* \in |\mathcal{O}G^*(B^*)| \subset |\mathcal{O}G(B)|$. By Proposition 18a there exists points $\tilde{x}$ arbitrarily close to x such that $g^m(\tilde{x}) = v^*$ for some positive integer m. For $\xi \in X_g$ proceed similarly using Proposition 18b. □

20 Corollary. *Let B^* be a G^* basic set with B the associated G basic set. If $x \in |\mathcal{O}G(B)|$ and there exists $\mathbf{z}^* \in K_{G^*}^{*+}$ with endset B^* such that $g^i(x) \in z_i^*$ for all $i \in \mathbf{Z}_+$ then x is a nonwandering point for g.*

Proof. For every neighborhood U of x and positive integer N there exists, by Proposition 18a, $\tilde{x} \in U$ and $m > N$ such that $g^m(\tilde{x}) = x$. □

On the other hand, when the center is a proper subset of X, there are always wandering points (see e.g. Akin (1993) Proposition 4.17). Some are easy to identify using the following.

21 Lemma. *Let $g : K^* \to K$ be a simplicial dynamical system on X with $d =$ dimension X.*

$$\{x \in X : g^i(x) \in (z^*)^\circ \text{ for some degenerate}$$

(4.53) simplex z^* of dimension d and some $i \in X\}$

is an open subset of X *consisting of wandering points.*

Proof. If z^* is degenerate then $g(z^*) \subset |\overline{S}^{d-1}(K)|$ which is a closed + invariant subset of X. If dim $z^* = d$ then $g^{-i}(z^*)^\circ$ is an open subset of X disjoint from $|\overline{S}^{d-1}(K)|$ for all $i \in \mathbf{Z}_+$. So if both assumptions hold then $g^j(g^{-i}(z^*)^\circ)$ is disjoint from $g^{-i}(z^*)^\circ$ for all $j > i$. Thus, the points of $g^{-i}(z^*)^\circ$ are wandering. □

To compute the individual basic set images we apply a variation of Barnsley's *iterated function system* methods from Barnsley (1988). For a compact metric space Y we let $C(Y)$ denote the space of closed subsets of Y with $C'(Y)$ the subspace of nonempty closed subsets. The *Hausdorff metric* is defined by

$$(4.54) \qquad d(A,B) = inf\{\epsilon \geq 0 : A \subset \overline{V}_\epsilon(B) \text{ and } B \subset \overline{V}_\epsilon(A)\}.$$

Equipped with this metric $C(Y)$ is compact and the empty set, $\emptyset$, is an isolated point. So $C'(Y)$ is compact as well (see, e.g. Akin (1993) Chapter 7). If $f : X \to Y$ is a continuous map of compact metric spaces we obtain the induced map $f : C(X) \to C(Y)$ by $A \mapsto f(A)$. It is easy to check that if f has Lipschitz constant L on X, i.e. $d(f(x), f(y)) \leq Ld(x,y)$, then it has Lipschitz constant L on $C'(X)$ as well. In fact, if $f_1, \ldots, f_k : X \to Y$ are maps with Lipschitz constant L, then defining $f_{1\ldots k} : C(X) \to C(Y)$ by $A \mapsto \cup_{i=1}^k f_i(A)$, we have

$$(4.55) \qquad d(f_{1\ldots k}(A), f_{1\ldots k}(B)) \leq Ld(A,B)$$

for $A, B \in C'(X)$ (see, e.g., Akin(1993) pp. 217-218). Now we apply this to our simplicial dynamical system $g : K^* \to K$.

On each simplex z of K we use the metric d_K on X to define the Hausdorff metric on $C(z)$ and $C'(z) = C(z)\backslash\{\emptyset\}$. If z^* is an expansion simplex of K^* with $z_1 = j(z^*)$ and $z_2 = g(z^*)$ we can compose the inverse of $g : z^* \to z_2$, i.e. $\overline{g}_{z^*}$ of (2.24), with the inclusion of z^* into z_1 and thus define $\overline{g}_{z^*} : z_2 \to z_1$. The induced map $\overline{g}_{z^*} : C(z_2) \to C(z_1)$ is given by

$$(4.56) \qquad \overline{g}_{z^*}(A) = g^{-1}(A) \cap z^* \subset z_1$$

for $A \subset z_2$.

Now let B^* be a basic set for G^* with B the associated G basic set so that $j(B^*) = B = g(B^*)$ and $B^* = j^{-1}(B) \cap g^{-1}(B)$ by Proposition 4d. We will denote by α a typical element of the product $\Pi_{z \in B} C(z)$ so that α_z is a closed subset of z for every $z \in B$.

22 Theorem. *Let B^* be a basic set for G^* with B the associated basic set for G. On $\Pi_{z \in B} C(z)$ define the continuous map $\overline{g}_{B^*}$ by*

$$\overline{g}_{B^*}(\alpha)_{z_1} = \cup \{ \overline{g}_{z^*}(\alpha_{z_2}) : z_2 \in B, z^* \in K^*, j(z^*) = z_1, g(z^*) = z_2 \}$$

$$= \cup \{ g^{-1}(\alpha_{g(z^*)}) \cap z^* : z^* \in B^*, z^* \subset z_1 \}, \tag{4.57}$$

for $z_1 \in B$.

The subset $\Pi_{z \in B} C'(z)$ is closed and + invariant for $\overline{g}_{B^*}$ and contains a unique attracting fixed point α^*. That is, for any $\alpha \in \Pi_{z \in B} C'(z)$

$$\mathrm{Lim}_{i \to \infty} d(\overline{g}^i_{B^*}(\alpha)_z, \alpha^*_z) = 0 \tag{4.58}$$

for all $z \in B$. Furthermore, the basic set image in X is given by

$$h^+(B^{*+}_{G^*}) = \cup_{z \in B} \alpha^*_z \subset X. \tag{4.59}$$

The basic set image in X_g is given by

$$h(B^*_{G^*}) = \{ \xi \in X_g : \xi_i \in h^+(B^{*+}_{G^*}) \text{ for all } i \in \mathbf{Z} \}. \tag{4.60}$$

Proof. Assume $\alpha_{z_2} \neq \emptyset$ for all $z_2 \in B$. For $z_1 \in B$ there exists $z_2 \in B \cap G(z_1)$ because B is a G basic set. So there exists $z^* \in B^*$ with $j(z^*) = z_1$ and $g(z^*) = z_2$. Thus, $\overline{g}_{B^*}(\alpha)_{z_1}$ contains the nonempty set $\overline{g}_{z^*}(\alpha_{z_2})$. This shows that $\Pi_{z \in B} C'(z)$ is + invariant for $\overline{g}_{B^*}$. On this set $\overline{g}_{B^*}$ is essentially a contraction mapping. In fact, from (4.55) and (2.56) we get the explicit estimate, for $\alpha, \beta \in \Pi_{z \in B} C'(z)$

$$\max_{z_1 \in B} d(\overline{g}_{B^*}(\alpha)_{z_1}, \overline{g}_{B^*}(\beta)_{z_1}) \leq$$

$$(1 - \frac{\theta}{d}) \max_{z_2 \in B} d(\alpha_{z_2}, \beta_{z_2}), \tag{4.61}$$

where d is the dimension of X and θ is the separation constant of K^* in K.

The existence of the unique attracting fixed point is the Banach Fixed Point Theorem. To identify the fixed point we use a typical example of Barnsley's monotonicity arguments. Notice that for $\alpha, \beta \in \Pi_{z\in B}C(z)$

$$\alpha_z \subset \beta_z \text{ for all } z \in B \Rightarrow$$

$$\overline{g}_{B^*}(\alpha)_z \subset \overline{g}_{B^*}(\beta)_z \text{ for all } z \in B. \tag{4.62}$$

Consequently, if $\alpha_z \supset \overline{g}_{B^*}(\alpha)_z$ for all $z \in B$ it follows that for each $z \in B$, $\{\overline{g}^m_{B^*}(\alpha)_z\}$ is a decreasing sequence of subsets of z. The intersection is the limit and so

$$\alpha_z \supset \overline{g}_{B^*}(\alpha)_z \text{ and } \alpha_z \neq \emptyset \text{ for all } z \in B \Rightarrow$$

$$\alpha^*_z = \cap^{\infty}_{m=0}\overline{g}^m_{B^*}(\alpha)_z \text{ for all } z \in B. \tag{4.63}$$

Begin with the maximal element α^0 defined by

$$\alpha^0_z = z \text{ for all } z \in B. \tag{4.64}$$

Now observe that for any $\alpha \in \prod_{z\in B} C(z)$, $z \in B$ and $x \in X$:

$$x \in (\overline{g}_{B^*})^m(\alpha)_z \Leftrightarrow$$

There exists a G^* chain $\{z^*_0, \ldots, z^*_{m-1}\}$ in B^* such that

$$z = j(z^*_0),\ g^i(x) \in z^*_i\ (i = 0, \ldots, m-1),\ g^m(x) \in \alpha_{g(z^*_{m-1})}. \tag{4.65}$$

With $m = 1$ this is the definition of $\overline{g}_{B^*}(\alpha)_{z_1}$ and for larger m it follows by induction.

With $\alpha = \alpha^0$ the condition $g^m(x) \in \alpha_{g(z^*_{m-1})}$ says $g^m(x) \in |B|$..

We now apply (4.63) with $\alpha = \alpha^0$ to prove that

$$x \in \alpha^*_z \Leftrightarrow$$

There exists $\mathbf{z}^* \in B^{*+}_{G^*}$ such that

$$z = j(z^*_0) \text{ and } x = h^+(\mathbf{z}^*). \tag{4.66}$$

The latter condition clearly implies $x \in (\overline{g}_{B^*})^m(\alpha^0_z)$ for all m by (4.65) and so $x \in \alpha^*_z$ by (4.63) applied to α^0. On the other hand, if $x \in (\overline{g}_{B^*})^m(\alpha^0)_z$ then by extending the G^* chain in B^* obtained from (4.65) we get $\mathbf{z}^{*\mathbf{m}} \in$

$B^{*+}_{G^*}$ such that $z = j(z_0^{*m})$ and $g^i(x) \in z_i^{*m}$ for $i \le m$. So if $x \in \alpha^*_z$ we can choose $\mathbf{z}^*$ a limit point by the sequence $\{\mathbf{z}^{*\mathbf{m}}\}$ to satisfy the condition of (4.66).

Equation (4.60) follows from Lemma 2.

Remark. We can use monotonicity in the other direction as well:

$$\alpha_z \subset \overline{g}_{B^*}(\alpha)_z \text{ and } \alpha_z \neq \emptyset \text{ for all } z \in B \Rightarrow$$

$$\alpha^*_z = \overline{\bigcup_{m=0}^{\infty} \overline{g}^m_{B^*}(\alpha)_z} \text{ for all } z \in B. \tag{4.67}$$

□

By using Barnsley type arguments we can sometimes describe the connectedness of $h^+(B^{*+}_{G^*})$.

23 Proposition. *Let B^* be a G^* basic set with B the associated G basic set. Let α^* be the attracting fixed point for the map $\overline{g}_{B^*}$ on $\Pi_{z \in B} C'(z)$.*

(a) Assume that for every $z \in B$ distinct elements z_1^*, z_2^* of B^* which are contained in z are disjoint. For every $z \in B$ the restriction of h^+ to $\{\mathbf{z}^* \in B^{*+}_{G^*} : \mathbf{z}^*_0 \subset z\}$ is a homeomorphism onto $\alpha^*_z \subset z$. The basic set image $h^+(B^{*+}_{G^*})$ is zero-dimensional.

(b) Assume that there exists $\alpha \in \Pi_{z \in B} C'(z)$ such that for every $z \in B$ α_z satisfies the conditions:

(1) α_z is connected.

(2) $\alpha_z \subset \alpha^*_z$.

(3) For every $z^* \in B^*$ such that $z^* \subset z$, $\alpha_z \cap \overline{g}_{z^*}(\alpha^*_{g(z^*)}) \neq \emptyset$.

Then for every $z \in B$ the set α^*_z is connected.

(c) Assume that there exists $\alpha \in \Pi_{z \in B} C'(z)$ such that for every $z \in B$ the following conditions hold:

(1) $\alpha_z \subset \overline{g}_{B^*}(\alpha)_z$.

(2) $\overline{g}_{B^*}(\alpha)_z$ is connected.

Then for every $z \in B$ the set α^*_z is connected.

Proof. (a) By (4.66) it is always true that h^+ maps $\{\mathbf{z}^* \in B^{*+}_{G^*} : \mathbf{z}^*_0 \subset z\}$ onto α^*_z. Now let $\mathbf{z}^*$ and $\tilde{\mathbf{z}}^*$ be distinct points of $B^{*+}_{G^*}$ with $j(\mathbf{z}^*_0) = j(\tilde{\mathbf{z}}^*_0) = z$.

Let k be the smallest $i \in \mathbf{Z}_+$ such that $\mathbf{z}_i^* \neq \tilde{\mathbf{z}}_i^*$. $\mathbf{z}_k^*, \tilde{\mathbf{z}}_k^*$ are distinct elements of B^* both contained in the same element of B, namely z if $k = 0$, otherwise in $g(\mathbf{z}_{k-1}^*) = g(\tilde{\mathbf{z}}_{k-1}^*) \in B$. So by hypothesis $\mathbf{z}_k^*$ and $\tilde{\mathbf{z}}_k^*$ are disjoint. With $x = h^+(\mathbf{z}^*)$ and $\tilde{x} = h^+(\tilde{\mathbf{z}}^*)$ we see that $x \neq \tilde{x}$ because $g^k(x)$ and $g^k(\tilde{x})$ lie in disjoint sets. This proves that h^+ is injective on the subset $\{\mathbf{z}^* \in B_{G^*}^{*+} : \mathbf{z}_0^* \subset z\}$. Thus the image is zero-dimensional because $B_{G^*}^{*+}$ is. Letting z vary over B we see that $h^+(B_{G^*}^{*+})$ is the finite union of compact zero-dimensional sets and so is zero-dimensional.

(b) Let $\alpha_z^0 = z$ which is connected for all $z \in B$. We prove inductively that $g^m(\alpha^0)_z$ is connected. Since the sequence of connected compacta $\{g^m(\alpha^0)_z\}$ decreases to α_z^* by (4.63) it will follow that α_z^* is connected. Let $A_0 = \alpha_z$ and for each $z^* \in B^*$ let $A_{z^*} = \overline{g}_{z^*}(\overline{g}_{B^*}^m(\alpha^0)_{g(z^*)})$. By definition $\overline{g}_{B^*}^{m+1}(\alpha^0)_z = \cup_{z^* \in B^*} A_{z^*}$. Because $\overline{g}_{z^*}$ is continuous and $\overline{g}_{B^*}^m(\alpha^0)_{g(z^*)}$ is connected by induction hypothesis, A_{z^*} is connected. A_0 is connected by assumption (1). By assumption (2), $A_0 \subset \alpha_z^* \subset \overline{g}_{B^*}^{m+1}(\alpha^0)_z$. Hence, $\overline{g}_{B^*}^{m+1}(\alpha^0)_z$ is $A_0 \cup \bigcup_{z^* \in B^*} A_{z^*}$. By assumption (3) each A_{z^*} meets A_0. Hence the union is connected. This completes the inductive step.

(c) Let $\tilde{\alpha} = \overline{g}_{B^*}(\alpha)$. We will prove that $\tilde{\alpha}$ satisfies the conditions of part (b). Assumption b(1) follows from assumption c(2). Assumption b(2) follows from c(1) together with (4.67). For every $z^* \in B^*$, $\alpha_{g(z^*)} \neq \emptyset$ since $\alpha \in \Pi_{z \in B} C'(z)$ and so

$$(4.68) \qquad \tilde{\alpha}_z \cap \overline{g}_{z^*}(\alpha_{g(z^*)}^*) \supset \overline{g}_{z^*}(\alpha_{g(z^*)})$$

implies assumption b(3) for $\tilde{\alpha}$. The conclusion of (c) then follows from (b). □

We conclude by applying these results to an important special case.

For any closed relation F on a compact metric space a basic set B is called *terminal* if it is minimal for the partial order $\mathcal{C}F$ induces upon the space of basic sets. That is, if for any basic set B_1, $B_1 \cap \mathcal{C}F(B) \neq \emptyset$ implies $B = B_1$.

In the special case at hand, let B^* be a k skeleton basic set for G^* with B the associated k skeleton basic set for G. As the proof of Proposition 18 shows there exists $\mathbf{z}^* \in K_{G^*}^{*+}$ with $\mathbf{z}_0^* \in B^*$ and $\mathbf{z}_i^*$ eventually in some cycle of vertices in $V(K)$. So if B^* is terminal for G^* it follows that $k = 0$. In general, we call B^* *terminal among k skeleton sets* when it is terminal for the restriction of G^* to $S^k(K^*)$, or equivalently, for any k skeleton basic set B_1^*, $\mathcal{O}G^*(B^*) \cap B_1^* \neq \emptyset$ implies $B^* = B_1^*$. Similarly, we call B terminal among k skeleton sets when it is terminal for the restriction of G to $S^k(K)$. Since distinct G^* basic sets have distinct associates it is easy to check that B^* is terminal among k skeleton sets iff its associate B is terminal among k skeleton sets.

24 Lemma. *Assume B^* is a basic set for G^*, terminal among k skeleton sets. Let $z^* \in \mathcal{O}G^*(B^*)$. The following conditions are equivalent:*

(1) $z^* \in B^*$.

(2) dim $z^* = k$ and $z^* \in |\mathcal{O}G^*|$, i.e. z^* is G^* periodic.

(3) For every $i \in \mathbf{Z}_+$, $g^i(z^*)\backslash|\overline{S}^{k-1}(K)| \neq \emptyset$.

Proof. (1) $\Rightarrow$ (2): Obvious.

(2) $\Rightarrow$ (3): Assume dim $z^* = k$ and for some positive integer i, $g^i(z^*) \subset |\overline{S}^{k-1}(K)|$. For $z_1^* \in K^*$ we have $z_1^* \subset g^i(z^*)$ iff $z_1^* \in (G^*)^i(z^*)$. So $g^i(z^*) \subset |\overline{S}^{k-1}(K)|$ implies any G^* chain of length at least i which begins with z^* terminates in $\overline{S}^{k-1}(K)$. Hence, z^* is not in any periodic G^* chain.

(3) $\Rightarrow$ (1): By (3) there exist arbitrarily long G^* chains beginning with z^* and remaining in $S^k(K^*)$. So by compactness there exists $\mathbf{z}^* \in S^k(K^*)^+_{G^*}$ with $\mathbf{z}_0^* = z^*$. Thus, the endset B_1^* for $\mathbf{z}^*$ is a k skeleton basic set. Because $B_1^* \subset \mathcal{O}G^*(z^*) \subset \mathcal{O}G^*(B^*)$ we have $B_1^* = B^*$ because B^* is terminal. Since B^* is an $\mathcal{O}G^* \cap \mathcal{O}G^{*-1}$ equivalence class it follows that $z_i^* \in B^*$ for all $i \in \mathbf{Z}_+$. In particular, $z^* \in B^*$. □

When g is nondegenerate, condition (3) holds whenever dim $z^* = k$. So if B^* is a basic set for G^* terminal among k skeleton sets and g is nondegenerate then $z^* \in B^*$ whenever dim $z^* = k$ and $z^* \in \mathcal{O}G^*(B^*)$. In particular, if $z \in B$, the G basic set associated with B^*, and $z^* \subset z$, then $z^* \in \mathcal{O}G^*(B^*)$ and so dim $z^* = k$ implies $z^* \in B^*$.

25 Corollary. *Let B^* be a k skeleton basic set for G^* with B the associated k skeleton basic set for G.*

(a) The following conditions are equivalent. When they hold we call B^* a polyhedral *basic set for G^*.*

(1) $|B^*| = |B|$.

(2) $h^+_{G^*}(B^{*+}_{G^*}) = |B^*|$.

(3) For $z^* \in K^*$, $\dim z^* = k$ *and* $z^* \subset |B|$ *implies* $z^* \in B^*$.

If B^* *is a polyhedral skeleton basic set then it is an interior basic set and* h^+ *is an almost homeomorphism from* $B^{*+}_{G^*}$ *onto* $|B|$. *If* B^* *is polyhedral then it is terminal among* k *skeleton basic sets.*

(b) If B^* *is terminal among* k *skeleton basic sets and* $g : K^* \to K$ *is nondegenerate then* B^* *is a polyhedral basic set.*

Proof. (a) (1) $\Rightarrow$ (3): Let $x \in (z^*)^\circ$ so that $q^*(x) = z^*$. If $z^* \subset |B|$ then $x \in |B|$. Hence, by (1), there exists $z^*_1 \in B^*$ with $x \in z^*_1$. Hence $z^* = q^*(x) < z^*_1$. Then $k = \dim z^* \leq \dim z^*_1 = k$ implies $z^* = z^*_1$ and so $z^* \in B^*$.

(3) $\Rightarrow$ (2): With α^0 defined by (4.64) it is clear that $\overline{g}_{B^*}(\alpha^0)_z = \cup\{z^* : z^* \in B^* \text{ and } z^* \subset z\}$ for $z \in B$. By (3) this is $\cup\{z^* : \dim z^* = k \text{ and } z^* \subset z\} = \alpha^0_z$. That is, α^0 is the fixed point α^* of $\overline{g}_{B^*}$ and so by (4.59), $h^+(B^{*+}_{G^*}) = |B|$. Since $h^+(B^{*+}_{G^*}) \subset |B^*|$, (2) follows.

(2) $\Rightarrow$ (1): If $|B^*|$ is a proper subset of $|B|$ then there exists $z \in B$ such that $(z)^\circ \setminus |B^*|$ is nonempty and $z^* \in B^*$ such that $g(z^*) = z$. The nonempty subset $(g|z^*)^{-1}((z)^\circ \setminus |B^*|)$ consists of points of $|B^*|$ mapping to the complement of $|B^*|$ and so consists of points in $|B^*| \setminus h^+(B^{*+}_{G^*})$.

By Theorem 15a, (1) and (2) imply B^* is an interior basic set and h^+ is an almost homeomorphism of $B^{*+}_{G^*}$ onto $|B|$.

Finally, suppose B^* is polyhedral $z^*_1 \in B^*$ and $z^*_2 \in G^*(z^*_1)$ with $\dim z^*_2 = k$. By definition of G^*, $z^*_2 \subset z_2 = g(z^*_1)$. Because $z^*_1 \in B^*$, $z_2 \in B$ by (4.9). By condition (3) it follows that $z^*_2 \in B^*$. Consequently, B^* is terminal among k skeleton basic sets.

(b) If B^* is a basic set terminal among k skeleton basic sets and g is nondegenerate then Lemma 24 ((3) $\Rightarrow$ (1)) implies that B^* satisfies condition (3) of (a) and so B^* is polyhedral. □

If B^* is a G^* basic set which is not polyhedral then we call B^* a *tattered* basic set.

5. Invariant Measures

For invariant measures we will follow the notation of Akin (1993). In particular, for X a compact metric space let $P(X)$ denote the space of Borel probability measures on X, equipped with the topology of weak convergence. For $\mu \in P(X)$ the *support* of $|\mu|$, is the smallest closed subset of X of measure 1.

If $f : X_1 \to X_2$ is a Borel measurable map then the induced map $f_* : P(X_1) \to P(X_2)$ is defined by $\mu \mapsto f_*\mu$ where

$$f_*\mu(A) = \mu(f^{-1}(A)) \tag{5.1}$$

for every Borel subset A of X_2 or, equivalently,

$$\int_{X_2} u(y) f_*\mu(dy) = \int_{X_1} u \circ f(x)\mu(dx) \tag{5.2}$$

for every continuous real-valued function u on X_2.

If f is a continuous map on X then f_* is a continuous map on $P(X)$. It is easy to check that

$$|f_*\mu| = f(|\mu|), \tag{5.3}$$

see Proposition 8.2 of Akin (1993). We then call $\mu \in P(X)$ an *invariant measure* for f if it is a fixed point for f_*. We denote by

$$|f_*| = \{\mu \in P(X) : f_*\mu = \mu\} \tag{5.4}$$

the set of f invariant measures. By (5.3) the support of an invariant measure is a closed invariant subset.

1 Lemma. *(a) Let f_α be a continuous map on X_α ($\alpha = 1, 2$) and $h : X_1 \to X_2$ be measurable mapping f_1 to f_2, i.e.*

$$h \circ f_1 = f_2 \circ h. \tag{5.5}$$

The induced map $h_* : P(X_1) \to P(X_2)$ maps f_{1*} to f_{2*} and maps the set $|f_{1*}|$ into $|f_{2*}|$.

(b) For f a continuous map on X let X_f be the associated sample path space with shift homeomorphism s_f and projection $\pi_0 : X_f \to X$ mapping s_f to f. The induced map $\pi_{0*} : P(X_f) \to P(X)$ restricts to a homeomorphism of $|s_{f*}|$ onto $|f_*|$.

Proof. (a) $h_*(f_{1*}\mu) = (h \circ f_1)_*\mu = (f_2 \circ h)_*\mu = f_{2*}(h_*\mu)$. In particular, $\mu = f_{1*}\mu$ implies $h_*\mu = f_{2*}(h_*\mu)$.

(b) Since π_0 maps s_f to f, π_{0*} maps $|s_{f*}|$ into $|f_*|$. Given $\nu \in |f_*|$ we exhibit a unique measure μ in $|s_{f*}|$ such that $\pi_{0*}\mu = \nu$.

The general relation results of Miller and Akin (1998) (cf. Theorem 3.1) imply there exists $\mu \in |s_{f*}|$ such that $\pi_{0*}\mu = \nu$. It suffices to prove uniqueness.

Because f is a mapping on X, $\pi_0^+ : X_f^+ \to X$ is a homeomorphism mapping s_f^+ to f. So there is a unique measure μ^+ on X_f^+ such that $\pi_{0*}^+\mu^+ = \nu$ and so $\pi_0^+ \circ \pi_f^+ = \pi_0$ implies $\pi_{f*}^+\mu = \mu^+$ where $\pi_f^+ : X_f \to X_f^+$ is the canonical projection.

The measure μ is determined by its value on cylindrical sets for X_f. Such a cylinder set is of the form $(s_f)^i(\pi_f^+)^{-1}(C)$ for some cylinder set C for X_f^+. Because μ is s_f invariant and $\pi_{f*}^+\mu = \mu^+$ we have

$$\mu((s_f)^i(\pi_f^+)^{-1}(C)) = \mu((\pi_f^+)^{-1}(C)) = \mu^+(C). \tag{5.6}$$

That is, μ is determined by μ^+.

Remark. The extreme points of the compact convex set $|f_*|$ in $P(X)$ are the ergodic measures. Since the homeomorphism $\pi_{0*} : |s_{f*}| \to |f_*|$ is linear it follows that $\mu \in |s_{f*}|$ is ergodic iff its image $\pi_{0*}\mu \in |f_*|$ is ergodic.
□

We will call a triple (X, f, μ) a *measurable dynamical system* when f is a continuous map on a compact metric space and $\mu \in |f_*|$. A morphism $h : (X_1, f_1, \mu_1) \to (X_2, f_2, \mu_2)$ is a Borel measurable map from X_1 to X_2 such that $h \circ f_1 = f_2 \circ h$ a.e. (μ_1) and $h_*\mu_1 = \mu_2$. The morphism h is an isomorphism when there exists $\tilde{h} : (X_2, f_2, \mu_2) \to (X_1, f_1, \mu_1)$ such that $\tilde{h} \circ h = 1_{X_1}$ a.e. (μ_1) and $h \circ \tilde{h} = 1_{X_2}$ a.e. (μ_2).

Now we return to our usual situation, $g : K^* \to K$ is a simplicial dynamical system with g the associated p.l. map on $X = |K^*| = |K|$.

2 Theorem. *(a) The induced maps* $(\pi_{G^*}^+)_* : P(K_{G^*}^*) \to P(K_{G^*}^{*+})$ *and* $\pi_{0*} : P(X_g) \to P(X)$ *restrict to linear homeomorphisms* $(\pi_{G^*}^+)_* : |(s_{G^*})_*| \to |(s_{G^*}^+)_*|$ *and* $\pi_{0*} : |s_{g*}| \to |g_*|$ *respectively.*

(b) Let μ *be an ergodic invariant measure for* g. *There exists a unique* G^* *basic set* B^* *such that*

$$|\mu| \subset h^+(B_{G^*}^{*+}),$$

(5.7) $$\mu(h^+(B^{*+}_{G^*} \cap \mathrm{Int}^+_k)) = 1$$

where $k = \dim z^*$ *for all* $z^* \in B^*$. *In particular,* B^* *is an interior basic set for* G^*.

There exists a unique invariant measure ν *for* $s^+_{G^*}$ *such that*

$$|\nu| \subset B^{*+}_{G^*}, \text{ and}$$

(5.8) $$h^+_* \nu = \mu.$$

The measure ν *is ergodic for* $s^+_{G^*}$ *and satisfies*

(5.9) $$\nu(B^{*+}_{G^*} \cap \mathrm{Int}^+_k) = 1.$$

Furthermore, the measurable dynamical systems $(K^{*+}_{G^*}, s^+_{G^*}, \nu)$ *and* (X, g, μ) *are isomorphic.*

(c) Let B^* *be a* k *skeleton interior basic set for* G^* *and let* ν *be an ergodic invariant measure for* $s^+_{G^*}$ *such that*

(5.10) $$|\nu| = B^{*+}_{G^*}.$$

Then equation (5.9) above holds and $\mu = h^+_* \nu$ *is an ergodic invariant measure for* g *such that*

$$|\mu| = h^+(B^{*+}_{G^*}),$$

(5.11) $$\mu(h^+(B^{*+}_{G^*} \cap \mathrm{Int}^+_k)) = 1.$$

Furthermore, the measurable dynamical systems $(K^{*+}_{G^*}, s^*_{G^*}, \nu)$ *and* (X, g, μ) *are isomorphic.*

Proof. (a) The result for π_{0*} is a direct application of Lemma 1b. When we identify the sample space of the map $s^+_{G^*}$ with that of G^* as in diagram (1.16) then the result for $(\pi^+_{G^*})_*$ follows from Lemma 1b as well.

(b) Call x a *convergence point* for the measure μ if

(5.12) $$\mathrm{Lim}_{n \to \infty} \frac{1}{n} \sum_{i=0}^{n-1} \delta_{g^i(x)} = \mu$$

where δ_y is the point measure concentrated at y and the limit is taken in $P(X)$. Let $\mathrm{Con}(\mu)$ denote the set of convergence points for μ. It is easy to check that

$$(5.13) \qquad x \in \mathrm{Con}(\mu) \Rightarrow |\mu| \subset \omega g(x).$$

that is, if U is an open set and $\{i : g^i(x) \in U\}$ is finite then $x \in \mathrm{Con}(\mu)$ implies $\mu(U) = 0$ (see Akin (1993) Proposition 8.8a). Since $|\mu|$ is closed and invariant, it follows that

$$(5.14) \qquad x \in \mathrm{Con}(\mu) \cap |\mu| \Rightarrow |\mu| = \omega g(x).$$

The Birkhoff Ergodic Theorem says that for an ergodic measure μ, $x \in \mathrm{Con}(\mu)$ a.e. (μ). Consequently,

$$(5.15) \qquad \mu(\mathrm{Con}(\mu) \cap |\mu|) = 1.$$

In particular, $\mathrm{Con}(\mu) \cap |\mu|$ is dense in $|\mu|$. Since it is nonempty (5.14) implies $|\mu|$ is a transitive subset of X and

$$(5.16) \qquad \mathrm{Con}(\mu) \cap |\mu| \subset \mathrm{Trans}_{g||\mu|}.$$

By Corollary 4.14 applied to $A = |\mu|$, the endset B^* of the transitive points of $g||\mu|$ satisfies

$$|\mu| \subset h^+(B^{*+}_{G^*})$$

$$(5.17) \qquad \mathrm{Trans}_{g||\mu|} \subset h^+(B^{*+}_{G^*} \cap \mathrm{Int}^+_k)$$

from which (5.7) follows from (5.16). On the other hand, if $\mu(h^+(\tilde{B}^{*+}_{G^*} \cap \mathrm{Int}^+_k)) = 1$ then by (5.15) there exists an interior chain $\mathbf{z}^* \in \tilde{B}^{*+}_{G^*}$ such that $h^+(\mathbf{z}^*) \in \mathrm{Con}(\mu) \cap |\mu|$. $\tilde{B}^* = B^*$ by Corollary 4.14. Uniqueness follows.

By Corollary 4.11, the restriction $h^+ : B^{*+}_{G^*} \cap \mathrm{Int}^+_k \to h^+(B^{*+}_{G^*} \cap \mathrm{Int}^+_k)$ is a homeomorphism with inverse q^{*+} and by Theorem 4.15 a

$$(5.18) \qquad B^{*+}_{G^*} \cap \mathrm{Int}^+_k = (h^+|B^{*+}_{G^*})^{-1}(h^+(B^{*+}_{G^*} \cap \mathrm{Int}^+_k)).$$

Let $\nu = (q^{*+})_*\mu$ and extend the measure from the subset $B^{*+}_{G^*} \cap \mathrm{Int}^+_k$ to all of $K^{*+}_{G^*}$. Clearly, ν satisfies (5.8) and (5.9) and is the unique measure satisfying (5.8) because (5.18) holds. Extending the homeomorphisms h^+ and q^{*+} to get Borel measurable maps between $K^{*+}_{G^*}$ and X we obtain an isomorphism of measurable dynamical systems. Since ergodicity is an isomorphism invariant ν is ergodic.

(c) As in (b) the Ergodic Theorem implies

$$(5.19) \qquad \nu(\mathrm{Con}(\nu) \cap |\nu|) = 1,$$

and so $\mathrm{Con}(\nu) \cap |\nu|$ is dense in $|\nu| = B^{*+}_{G^*}$

$$\mathrm{Con}(\nu) \cap |\nu| \subset \mathrm{Trans}_{B^*} \tag{5.20}$$

where Trans_{B^*} denotes the set of transitive points of $s^+_{G^*}|B^{*+}_{G^*}$. Because B^* is assumed to be an interior basic set (4.44) holds:

$$\mathrm{Trans}_{B^*} \subset B^{*+}_{G^*} \cap \mathrm{Int}^+_k. \tag{5.21}$$

So (5.9) follows from (5.20) and (5.19). Then with $\mu = h^+_* \nu$ we obtain (5.11) from (5.3) and (5.18). Isomorphism and hence ergodicity follows as in (b). □

Where d is the dimension of X, define

$$\mathrm{Int}^+ = \cup^d_{k=0} \mathrm{Int}^+_k \subset K^{*+}_{G^*}. \tag{5.22}$$

By Proposition 4.10a ((1) $\Rightarrow$ (3)) the continuous map h^+ on Int^+ is injective with inverse map q^{*+} on

$$h^+(\mathrm{Int}^+) = \cup^d_{k=0} h^+(\mathrm{Int}^+_k) \subset X. \tag{5.23}$$

By Corollary 11a, q^{*+} is continuous on each $h^+(\mathrm{Int}^+_k)$ and so it is Borel measurable on $h^+(\mathrm{Int}^+)$.

3 Corollary. *The subset $h^+(\mathrm{Int}^+)$ is a G_δ subset of full measure with respect to g. That is, if μ is any g invariant measure on X then*

$$\mu(h^+(\mathrm{Int}^+)) = 1. \tag{5.24}$$

In addition, there is then a unique $s^+_{G^*}$ invariant measure ν on $K^{*+}_{G^*}$ satisfying

$$\nu(\mathrm{Int}^+) = 1, \text{ and}$$

$$h^+_* \nu = \mu. \tag{5.25}$$

The continuous mapping $h^+ : \mathrm{Int}^+ \to h^+(\mathrm{Int}^+)$ extends to an isomorphism of measurable dynamical systems $(K^{*+}_{G^*}, s^+_{G^*}, \nu) \to (X, g, \mu)$.

Proof. By Corollary 4.11a, each Int^+_k and its image are G_δ sets. So the finite unions Int^+ and $h^+(\mathrm{Int}^+)$ are G_δ's as well.

It is one of the consequences of the Ergodic Theorem that $\cup\{\mathrm{Con}(\mu) \cap |\mu| : \mu$ as ergodic measure for $g\}$ is a Borel set of full measure, see Akin (1993) Proposition 8.7 and 8.8. By (5.16) and (5.17) this union is contained in $h^+(\mathrm{Int}^+)$. So (5.24) follows. Then use (5.1) with the Borel map q^{*+} to define ν by extending the measure $(q^{*+})_*\mu$ from Int^+ to all of $K_{G^*}^{*+}$. Clearly ν is the unique measure on $K_{G^*}^{*+}$ satisfying (5.25). It is $s_{G^*}^+$ invariant because $s_{G^*}^+ \circ q^{*+} = q^{*+} \circ g$, see Lemma 1a. h^+ and q^{*+} extend to define inverse isomorphisms of the associated measurable dynamical systems. □

A useful application of these invariant measures is a variation of what Barnsley calls the *chaos game* applied to describe a basic set image $h^+(B_{G^*}^{*+})$.

Start with a point y of z, a simplex of B, the associated G basic set for B^*. Let $\delta(y) \in \Pi_{z\in B} C(z)$ with

$$\delta(y)_w = \begin{cases} \{y\} & w = z \\ \emptyset & w \neq z. \end{cases} \tag{5.26}$$

Now start iterating the set operator $\overline{g}_{B^*}$ of (4.57) with initial point $\delta(y)$. By (4.65) it is clear that after at most n steps, where n is the cardinality of B, we arrive at an element of $\Pi_{z\in B} C'(z)$. So by Theorem 4.22 the sequence $\{(\overline{g}_{B^*})^i(\delta(y))\}$ converges to the fixed point α^* as $i \to \infty$.

However, this procedure is extremely unwieldly because the number of points increases exponentially at each iterate.

Instead, Barnsley's procedure takes a random point $\mathbf{z}^*$ of $B_{G^*}^*$. Choose an arbitrary point y in $z_0 = g(z_{-1}^*)$ and define, inductively:

$$\eta_{-1} = \overline{g}_{z_{-1}^*}(y)$$

$$\eta_{i-1} = \overline{g}_{z_{i-1}^*}(\eta_i) \quad \text{for } i < 0. \tag{5.27}$$

That is, we just proceed backwards along the chain using $\overline{g}_{z_{i-1}^*}$ the inverse of the linear isomorphism of $g : z_{i-1}^* \to z_i = j(z_i^*)$. We thus have

$$\eta_i = g(\eta_{i-1}) \text{ and } \eta_i \in z_i^* \text{ for } i < 0. \tag{5.28}$$

Define:

$$\mathrm{Lim\,sup}_{i\to-\infty}\{\eta_i\} = \cap_{m=0}^{\infty}\overline{\{\eta_i : i < -m\}}. \tag{5.29}$$

It is always true that this Lim sup is contained in $h^+(B^{*+}_{G^*})$ with equality for almost every choice of $\mathbf{z}^*$.

4 Proposition. *Let B^* be a G^* basic set.*

(a) Let $\mathbf{z}^* \in B^*_{G^*}$ with $\xi = h(\mathbf{z}^*)$ so that $\{\xi_i\}$ is the unique g chain defined for $i \in \mathbf{Z}$ such that $\xi_i \in z^*_i$ for all $i \in \mathbf{Z}$. If $\{\eta_i\}$ is a g chain defined for negative integers i such that $\eta_i \in z^*_i$ for all $i < 0$ then

$$\text{Lim}_{i\to-\infty} d_K(\xi_i, \eta_i) = 0 \tag{5.30}$$

and

$$\text{Lim sup}_{i\to-\infty}\{\eta_i\} = \text{Lim sup}_{i\to-\infty}\{\xi_i\} =$$

$$= h^+(\pi^+_{G^*}(\alpha s_{G^*}(\mathbf{z}^*))). \tag{5.31}$$

Thus, Lim sup$_{i\to-\infty}\{\eta_i\}$ is a subset of $h^+(B^{*+}_{G^*})$ and it depends only on $\mathbf{z}^*$.

(b) The set $\{\mathbf{z}^* \in B^*_{G^*} : \alpha s_{G^*}(\mathbf{z}^*) = B^*_{G^*}\}$, the set of transitive points for the homeomorphism $(s_{G^*})^{-1}$ on $B^*_{G^*}$, is a G_δ subset of $B^*_{G^*}$. For every ergodic s_{G^*} invariant measure ν such that $|\nu| = B^*_{G^*}$

$$\nu(\{\mathbf{z}^* \in B^*_{G^*} : \alpha s_{G^*}(\mathbf{z}^*) = B^*_{G^*}\}) = 1 \tag{5.32}$$

and so for a.e. (ν) $\mathbf{z}^* \in B^*_{G^*}$ we have $h^+(\pi^+_{G^*}(\alpha s_{G^*}(\mathbf{z}^*))) = h^+(B^{*+}_{G^*})$.

Proof. (a) The limit in (5.30) follows from Theorem 2.11, since the simplices $z^*_i \in B^*$ are nondegenerate, and so

$$\eta_{i-1} = \overline{g}_{z^*_{i-1}}(\eta_i) \text{ and } \xi_{i-1} = \overline{g}_{z^*_{i-1}}(\xi_i) \quad \textit{for } i < 0. \tag{5.33}$$

The first equation of (5.31) clearly follows. Now $\xi_i = h^+(\pi^+_{G^*}((s_{G^*})^i(\mathbf{z}^*)))$ for all $i \in \mathbf{Z}$ and $\alpha s_{G^*}(\mathbf{z}^*)$ is Lim sup$_{i\to-\infty}\{(s_{G^*})^i(\mathbf{z}^*)\}$. So the compactness argument used to prove (1.28) yields the second equation in (5.31). Since $\mathbf{z}^*$ is a point of the closed, s_{G^*} invariant set $B^*_{G^*}$ it follows that $\alpha s_{G^*}(\mathbf{z}^*) \subset B^*_{G^*}$ and so $h^+(\pi^+_{G^*}(\alpha s_{G^*}(\mathbf{z}^*))) \subset h^+(B^{*+}_{G^*})$.

(b)Since $\alpha s_{G^*} = \omega(s^{-1}_{G^*})$ for the homeomorphism s_{G^*} we see that $\alpha s_{G^*}(\mathbf{z}^*) = B^*_{G^*}$ exactly when $\mathbf{z}^*$ is a transitive point for $(s_{G^*})^{-1}$ on $B^*_{G^*}$. The set of such transitive points is always G_δ and it is dense since B^* is a basic set for G^* and hence for $(G^*)^{-1}$. A measure ν is invariant for s_{G^*} iff it is invariant for the inverse and so the set of s_{G^*} ergodic measures, the extreme points of $|(s_{G^*})_*|$, is the same as the set of $(s_{G^*})^{-1}$ ergodic measures. Applying the Ergodic Theorem to the system $(B^*_{G^*}, (s_{G^*})^{-1}, \nu)$ the analogues of (5.19) and (5.20) yield (5.32). Finally, observe that $\alpha s_{G^*}(\mathbf{z}^*) = B^*_{G^*}$ implies $h^+(\pi^*_{G^*}(\alpha s_{G^*}(\mathbf{z}^*))) = h^+(B^{*+}_{G^*})$ because $\pi^+_{G^*}(B^*_{G^*}) = B^{*+}_{G^*}$. □

The practical way to choose and use the random point $\mathbf{z}^*$ is to generate the sequence $z^*_0, z^*_{-1}, z^*_{-2}, \ldots$ iteratively using a stochastic process. For example, fix for each $z \in B$ a distribution over the set $(g|B^*)^{-1}(z) = \{z^* \in B^* : g(z^*) = z\}$ and proceed from z^*_i to z^*_{i-1} by using the distribution associated with $z_i = j(z^*_i)$. This is a Markov process and leads us to our most natural examples of ergodic measures on $B^*_{G^*}$ and $B^{*+}_{G^*}$, the Markov measures. In order to describe them some matrix notation will be helpful.

Let I and J be finite sets. By an *IJ matrix* we mean a *nonnegative*, real-valued function M on the set $I \times J$ regarded as a matrix with rows indexed by I and columns indexed by J. As usual, we can multiply an LI matrix times an IJ matrix to get an LJ matrix where L is another finite set. We call an IJ matrix *square* when $I = J$ (same set, not just same cardinality).

For an IJ matrix, M, the *support* of M, $F(M)$, is a relation from J to I:

$$F(M) = \{(j, i) \in J \times I : M_{ij} > 0\}. \tag{5.34}$$

That is, we have

$$i \in F(M)(j) \Leftrightarrow M_{ij} > 0. \tag{5.35}$$

Notice the annoying reversal of order. This comes from the convention that relations map from first coordinate to second while matrices map from column space to row space. These conventions do associate matrix multiplication with relation composition. That is, if Q is an LI matrix and M is an IJ matrix then

$$F(QM) = F(Q) \circ F(M). \tag{5.36}$$

In the reverse direction, given a relation $F \subset J \times I$ we define the IJ *indicator matrix* $K(F)$ to be the 0/1 matrix with support F. That is,

$$K(F)_{ij} = \begin{cases} 1 & (j, i) \in F \\ 0 & (j, i) \notin F. \end{cases} \tag{5.37}$$

If $F : I \to L$ is a relation and $f : J \to I$ is a function then we have

$$K(F \circ f) = K(F)K(f). \tag{5.38}$$

If M is an II matrix, then the matrix powers M^k $k = 0, 1, 2, \dots$ are II matrices with M^0 defined to be the identity II matrix, i.e. $K(1_I)$. From (5.36) we have for $k = 0, 1, 2, \dots$

$$F(M^k) = F(M)^k \tag{5.39}$$

relating matrix powers to relation interates.

An II matrix M is called *irreducible* if for every $i, j \in I$, there exists $k > 0$ such that $M^k_{ij} > 0$. By (5.39) this says exactly that the entire set I is a single basic set for the relation $F(M)$, i.e. $\mathcal{O}F(M) = I \times I$.

Now assume that M is an irreducible II matrix. The Perron-Frobenius Theorem (Gantmacher (1959) Section XIII.2) says that γ, the maximum modulus of the eigenvalues of M, is itself a simple eigenvalue with positive left and right eigenvectors l, r. Regarding these as column vectors we have:

$$\begin{gathered} l_i, r_i > 0 \ \text{ for all } i. \\ l^T M = \gamma l^T \ \text{ and } \ Mr = \gamma r. \\ l^T r = \sum_i l_i r_i = 1. \end{gathered} \tag{5.40}$$

The last equation is a normalization. Define $\Delta(l)$ to be the diagonal II matrix with $\Delta(l)_{ii} = l_i$ and define:

$$\begin{gathered} P = \frac{1}{\gamma}\Delta(l)M\Delta(l)^{-1}, \text{ i.e.} \\ P_{ij} = \frac{l_i M_{ij}}{\gamma l_j} \ \text{ for } i, j \in I \end{gathered} \tag{5.41}$$

Define the vector p by

$$\begin{gathered} p = \Delta(l)r, \text{ i.e.} \\ p_i = l_i r_i \ \text{ for } i \in I. \end{gathered} \tag{5.42}$$

Clearly, P is an II matrix with the same support as M. In addition, P is a stochastic matrix. For each j the column sum is 1. The vector p is positive and

$$Pp = p \text{ with } \sum_i p_i = 1. \tag{5.43}$$

On the other hand, if M was stochastic then $l_i = 1$ for all $i \in I$ defines an eigenvector with eigenvalue $\gamma = 1$. Uniqueness in the Perron-Frobenius Theorem implies that γ is the maximum modulus eigenvalue. Then by (5.41) and (5.42), $P = M$ and $p = r$. So we call the stochastic matrix P constructed above the *stochastic retract* of M and p the associated *distribution vector.*

Now let $\overline{P}$ denote the stochastic retract of the transpose of M and $\overline{p}$ be the associated distribution vector. Since transposition interchanges l and r we have:

$$\overline{P}_{ij} = \frac{M_{ji}r_i}{\gamma r_j} = \frac{P_{ji}p_i}{p_j}$$

$$\overline{p}_i = p_i \text{ for } i, j \in I. \tag{5.44}$$

The stochastic matrix $\overline{P}$ is called the *reverse matrix* for P. If P describes a Markov process then $\overline{P}$ describes the reverse of the same process. That is, for $j_0, \ldots, j_n \in I$ we define the associated probability of the sequence to be:

$$\begin{aligned} p_{j_0 j_1 \cdots j_n} &\equiv \\ P_{j_n j_{n-1}} \cdots P_{j_1 j_0} p_{j_0} &= \\ \overline{P}_{j_0 j_1} \cdots \overline{P}_{j_{n-1} j_n} p_{j_n} \end{aligned} \tag{5.45}$$

5 Lemma. *Let P be a stochastic II matrix with support relation $F = F(P)$ on I. Let s_F be the shift homeomorphism on the sample path space I_F and s_F^+ the shift map on I_F^+. There is a unique measure ν^P on $I^{\mathbf{Z}}$ defined on cylinder sets by:*

$$\begin{aligned} \nu^P(\{\xi \in I^{\mathbf{Z}} : \xi_k = j_0, \xi_{k+1} = j_1, \ldots, \xi_{k+n} = j_n\}) \\ = p_{j_0 \cdots j_n} \end{aligned} \tag{5.46}$$

for all integers k and all sequences $j_0, \ldots, j_n$ in I. ν^P is an ergodic, shift invariant measure with

$$|v^P| = I_F \tag{5.47}$$

and its projection via $(\pi^+)_*$ *to* $I^{\mathbf{Z}_+}$, *also denoted* ν^P, *is an ergodic, shift invariant measure with*

$$|\nu^P| = I_F^+. \tag{5.48}$$

The measures ν^P *are called the* Markov measures *on* I_F *and* I_F^+ *associated with the stochastic matrix* P.

Proof. This construction is standard (see, e.g., Billingsley (1965) p. 30). The definition (5.45) yields a distribution vector on the product set I^{n+1}. Projecting by summation from the first or last coordinate yields the distribution vector on I^n. The coherent definitions yield a measure on $I^{\mathbf{Z}}$ satisfying (5.46). The resulting measure ν^P is clearly shift invariant. For the proof of ergodicity see Billingsley (1965), as above. By (5.39) it follows that $p_{j_0 \dots j_n} > 0$ iff $\{j_0, \dots, j_n\}$ is an F chain. Because the cylindrical sets form a basis for the topology of $I^{\mathbf{Z}}$ the support result (5.47) follows. Since $\pi_F^+(I_F) = I_F^+$, (5.48) follows from (5.3). □

When P is the stochastic retract of the matrix M then from (5.41) it follows that

$$\begin{aligned} -\sum_{i,j} P_{ij} p_j \ln(l_i/\gamma l_j) &= \\ -\sum_{i,j} \frac{l_i M_{ij} r_j}{\gamma} [(\ln l_i) - (\ln l_j) - \ln \gamma] & \\ = \ln \gamma. \end{aligned} \tag{5.49}$$

In particular, if M is a 0/1 matrix then

$$-\sum_{ij} P_{ij} p_j \ln(P_{ij}) = \ln \gamma. \tag{5.50}$$

The expression on the left is the entropy of the shifts s_F and s_F^+ with respect to the Markov measures ν^P (see Billingsley (1965) equation (7.3)).

For a relation F on I the measures ν^P associated with the stochastic retract of the characteristic matrix $K(F)$ are called the *Parry measures* for s_F and s_F^+. So (5.50) computes the entropy of the Parry measures.

Now we apply this to the case of a simplicial dynamical system $g : K^* \to K$ with associated relations $G^* = J^{-1} \circ g$ on K^* and $G = g \circ J^{-1}$ on K. Let B^* be a basic set for G^* and B be the associated basic set for G. Recall from Proposition 4.4 that the restricted relation $J \cap B^* \times B$ is the surjective function $j_{B^*} : B^* \to B$; g restricts to a surjective function $g_{B^*} : B^* \to B$; and $B^* = \{z^* \in K^* : j(z^*), g(z^*) \in B\}$. For the restriction $G^*_{B^*}$ of G^* to B^*, the characteristic matrix satisfies, by (5.38):

$$K(G^*_{B^*}) = K(j^{-1}_{B^*})K(g_{B^*}) = K(j_{B^*})^T K(g_{B^*})$$

$$K(G^{*-1}_{B^*}) = K(G^*_{B^*})^T = K(g^{-1}_{B^*})K(j_{B^*}). \tag{5.51}$$

For G_B define

$$M(G_B) = K(g_{B^*})K(j^{-1}_{B^*}) = K(g_{B^*})K(j_{B^*})^T$$

$$M(G^{-1}_B) = M(G_B)^T = K(j_{B^*})K(g^{-1}_{B^*}). \tag{5.52}$$

Because $j^{-1}_{B^*}$ is not a function (5.38) does not apply and $M(G_B)$ is not the characteristic matrix of G_B. In fact, for $z_1, z_2 \in B$:

$$M(G_B)_{z_2 z_1} = \#\{z^* \in B^* : j(z^*) = z_1 \text{ and } g(z^*) = z_2\} \tag{5.53}$$

where # denotes the cardinality of the set. Notice that for $z^* \in K^*$. $j(z^*) = z_1$ and $g(z^*) = z_2$ implies $z^* \in B^*$ by (4.10). We do have that the support of $M(G_B)$ is G_B, i.e.

$$F(M(G_B)) = G_B \tag{5.54}$$

by definition of the relation G.

Since B^* and B are basic sets the matrices $K(G^*_{B^*})$ and $M(G_B)$ are irreducible. Let l and r be the positive left and right eigenvectors for $M(G_B)$ with positive eigenvalue γ. Thus,

$$l^T M(G_B) = l^T K(g_{B^*})K(j^{-1}_{B^*}) = \gamma l^T$$

$$M(G_B) r = K(g_{B^*})K(j^{-1}_{B^*}) r = \gamma r. \tag{5.55}$$

Now define vectors l^* and r^* on B^* by:

$$l^{*T} = l^T K(g_{B^*}) \text{ and } r^* = \frac{1}{\gamma} K(j^{-1}_{B^*}) r. \tag{5.56}$$

Multiplying the first equation on the right by $K(j^{-1}_{B^*})$ and the second on the left by $K(g_{B^*})$ we obtain from (5.55)

$$l^{*T} K(j^{-1}_{B^*}) = \gamma l^T \text{ and } K(g_{B^*}) r^* = r. \tag{5.57}$$

From which it follows that l^* and r^* are positive eigenvalues for $K(G_{B^*})$ associated with the positive eigenvalue γ:

$$l^{*T}K(G^*_{B*}) = l^{*T}K(j_{B^*}^{-1})K(g_{B^*}) = \gamma l^{*T}$$

$$K(G^*_{B^*})r^* = K(j_{B^*}^{-1})K(g_{B^*})r^* = \gamma r^*. \tag{5.58}$$

In fact, we could just as easily have reversed the process. However, since B usually has a smaller cardinality than B^*, l and r are easier to compute. By (5.56) we have for $z^* \in B^*$:

$$l^*_{z^*} = l_{g(z^*)} \text{ and } r^*_{z^*} = r_{j(z^*)}/\gamma. \tag{5.59}$$

The equations (5.57) say for $z \in V$:

$$\gamma l_z = \sum\{l_{g(z^*)} : z^* \in B^* \text{ with } j(z^*) = z\}$$

$$\gamma r_z = \sum\{r_{j(z^*)} : z^* \in B^* \text{ with } g(z^*) = z\}. \tag{5.60}$$

Recall that we have normalized l and r so that $l^T r = 1$. So from (5.56) and (5.57)

$$l^{*T}r^* = l^T K(g_{B^*})r^* = l^T r = 1. \tag{5.61}$$

The associated distribution vectors p on B and p^* on B^* are given by

$$p_z = l_z r_z \text{ and } p^*_{z^*} = l^*_{z^*}r^*_{z^*} = l_{g(z^*)}r_{j(z^*)}/\gamma. \tag{5.62}$$

Because g_{B^*} and j_{B^*} are maps it follows from (5.59) that

$$\Delta(l)K(g_{B^*}) = K(g_{B^*})\Delta(l^*),$$

$$\Delta(r)K(j_{B^*}) = \gamma K(j_{B^*})\Delta(r^*). \tag{5.63}$$

It follows that we can write the stochastic retract P^* of $K(G^*_{B^*})$ as follows (cf. (5.41))

$$P^* = \frac{1}{\gamma}\Delta(l^*)K(G_{B^*})\Delta(l^*)^{-1}$$

$$= \frac{1}{\gamma}\Delta(l^*)K(j_{B^*}^{-1})K(g_{B^*})\Delta(l^*)^{-1} = Q^*K(g_{B^*})$$

$$\text{where } Q^* = \frac{1}{\gamma}\Delta(l^*)K(j_{B^*}^{-1})\Delta(l)^{-1}. \tag{5.64}$$

The reverse matrix $\overline{P}^*$, i.e. the stochastic retract of the transpose can be similarly written

$$\overline{P}^* = \frac{1}{\gamma}\Delta(r^*)K(G_{B^*}^*)^{-1}\Delta(r^*)^{-1} =$$

$$= \frac{1}{\gamma}\Delta(r^*)K(g_{B^*}^{-1})K(j_{B^*})\Delta(r^*)^{-1} = \overline{Q}^*K(j_{B^*})$$

$$\text{where } \overline{Q}^* = \Delta(r^*)K(g_{B^*}^{-1})\Delta(r)^{-1}. \tag{5.65}$$

P^* and $\overline{P}^*$ are B^*B^* stochastic matrices. Q^* and $\overline{Q}^*$ are B^*B stochastic matrices with:

$$Q^*_{z^*z} = \begin{cases} \frac{l_{g(z^*)}}{\gamma l_z} & \text{if } j(z^*) = z \\ 0 & \text{otherwise} \end{cases}$$

$$\overline{Q}^*_{z^*z} = \begin{cases} \frac{r_{j(z^*)}}{\gamma r_z} & \text{if } g(z^*) = z \\ 0 & \text{otherwise} \end{cases} \tag{5.66}$$

If we write u for the B vector with $u_z = 1$ for all $z \in B$ then by (5.55)

$$l^T M(G_B)u = \gamma l^T u$$

$$u^T M(G_B)r = \gamma u^T r. \tag{5.67}$$

From the equation (5.53) we see that this says:

$$\frac{\sum \#g_{B^*}^{-1}(z)l_z}{\sum l_z} = \gamma = \frac{\sum \#j_{B^*}^{-1}(z)r_z}{\sum r_z} \tag{5.68}$$

where the sums are taken over all $z \in B$ and, e.g., $\#g_{B^*}^{-1}(z)$ is the cardinality of $\{z^* \in B^* : g(z^*) = z\}$. Thus the eigenvalue γ is a weighted average of each of these numerical functions. It thus represents something like the "degree" of the p.l. map $g : |B^*| \to |B|$. For example, if $\#g_{B^*}^{-1}(z)$ is independent of $z \in B$ then with γ this common value, then $M(G_B)u = \gamma u$ and so r is a multiple of u. It then follows from (5.66) that for each $z \in B$, the distribution $\overline{Q}^*_{z^*z}$ on the set $g_{B^*}^{-1}(z)$ is uniform with value $1/\gamma$ at each z^*. Similarly, if $\#j_{B^*}^{-1}(z)$ is independent of $z \in B$ then l is a multiple of u

and for each $z \in B$ the forward distribution Q_{z^*z} on $j_{B^*}^{-1}(z)$ is uniform with value $1/\gamma$ at each z^*.

It is now clear how the Markov process associated with P^* and $\overline{P}^*$ generates G^* chains. Begin with z_0^* chosen randomly using the distribution p^* on B^*. Let $z_1 = g(z_0^*)$. Choose z_1^* randomly among the set of $z^* \in B^*$ such $j(z) = z_1$ by using the distribution $Q^*_{z^*z_1}$. Let $z_2 = g(z_1^*)$, choose z_2^* using $Q^*_{z^*z_2}$ and continue inductively. For the negative direction let $z_0 = j(z_0^*)$. Choose z_{-1}^* randomly among the set of z^* in B^* such that $g(z^*) = z_0$, using the distribution $\overline{Q}^*_{z^*z_0}$. The associated measures ν^{P^*} on $B^*_{G^*}$ and $B^{*+}_{G^*}$ are the Markovian Parry measures for the relation G^* on B^*.

6 Theorem. *For a simplicial dynamical system $g : K^* \to K$ let B^* be an interior k skeleton basic set and ν^{P^*} denote the Parry measures on $B^*_{G^*}$ and $B^{*+}_{G^*}$, invariant with respect to the shifts s_{G^*} and $s^+_{G^*}$, respectively. Denote by μ_{B^*} the measure $h^+_*(\nu^{P^*})$ induced by the map $h^+ : B^{*+}_{G^*} \to h^+(B^{*+}_{G^*})$. The continuous map h^+ defines an isomorphism from the measurable dynamical system $(B^{*+}_{G^*}, s^+_{G^*}, \nu^{P^*})$ to $(h^+(B^{*+}_{G^*}), g, \mu_{B^*})$. The entropy of each of these measurable dynamical systems, as well as the topological entropy of $s^+_{G^*}$ on $B^{*+}_{G^*}$ and g on $h^+(B^{*+}_{G^*})$, is $ln\gamma(B^*)$ where $\gamma(B^*)$ is the positive eigenvalue of the characteristic matrix $K(G^*_{B^*})$.*

Proof. The Markov measures ν^{P^*} are constructed using P^* and $\overline{P}^*$ as in Lemma 5. By (5.50) and Billingsley (1965) equation (7.3) $\ln \gamma$ is the entropy of the Markov shifts $(B^*_{G^*}, s_{G^*}, \nu^{P^*})$ and $(B^{*+}_{G^*}, s^+_{G^*}, \nu^{P^*})$. By (5.48) we can apply Theorem 2c to ν^{P^*} so that $h^+ : (B^{*+}_{G^*}, s^+_{G^*}, \nu^{P^*}) \to (h^+(B^{*+}_{G^*}), g, \mu_{B^*})$ is an isomorphism of measurable dynamical systems. Because entropy is an isomorphism invariant it follows that $\ln \gamma$ is the entropy of the measurable system $(h^+(B^{*+}_{G^*}), g, \mu_{B^*})$ as well.

For topological entropy we refer to Robinson (1995) Section 8.1. In particular, in Theorem 8.1.9, Robinson computes the topological entropy for the subshift of finite type $s^+_{G^*}$ on $B^{*+}_{G^*}$ and he shows it is $\ln \gamma$. Because the system g on $h^+(B^{*+}_{G^*})$ is a factor of this subshift, i.e. $h^+ : B^{*+}_{G^*} \to h^+(B^{*+}_{G^*})$ is a continuous surjection mapping $s^+_{G^*}$ to g, it follows from Theorem 8.1.7 of Robinson (1995) that the topological entropy of g on $h^+(B^{*+}_{G^*})$ is bounded by that of $s^+_{G^*}$ on $B^{*+}_{G^*}$, i.e. by $\ln \gamma$.

Finally, we apply the result of Misiurewicz (1976) that the topological entropy is the supremum of the measure theoretic entropies obtained by using all invariant measures. Since $\ln \gamma$ is the entropy of $(h^+(B^{*+}_{G^*}), g, \mu_{B^*})$ it follows that $\ln \gamma$ equals the topological entropy of g on $h^*(B^{*+}_{G^*})$. Actually, we only need the original result of Goodwyn (1972) that topological entropy is an upper bound for all the measure theoretic entropies. □

7 Corollary. *For a simplicial dynamical system $g : K^* \to K$, the topological entropy of the piecewise linear map g on $X = |K^*| = |K|$ is*

$$\max\{ln\gamma(B^*) : B^* \text{ is an interior basic set for } G^*\}, \tag{5.69}$$

where $\gamma(B^*)$ is the positive eigenvalue of the characteristic matrix $K(G^*_{B^*})$.

Proof. By Theorem 4.15b the Birkhoff center, the closure of the set of recurrent points, is $\cup\{h^+(B^{*+}_{G^*}) : B^*$ an interior basic set for $G^*\}$. By Robinson (1995) Theorem 8.1.3 the entropy of g on a finite union of closed invariant subsets is just the maximum of the entropies on each piece. So Theorem 6 implies that expression (5.69) is the entropy of the restriction of g to its center.

Finally, the entropy of g is equal to the entropy of its restriction to the center. To see this one can observe that the support of every invariant measure is contained in the center (see Akin (1993) Proposition 4.17 and equation (8.22)) and then apply the Misiurewicz (1976) result that topological entropy is the sup of the measure theoretic entropies. □

The other natural class of examples comes from Lebesgue measure. Recall that if λ is Lebesgue measure on $\mathbf{R}^n$ and $T : \mathbf{R}^n \to \mathbf{R}^n$ is a linear isomorphism then

$$|\det(T)| T_* \lambda = \lambda. \tag{5.70}$$

For example, the image of the unit hypercube with sides the unit vectors $e_1, \dots, e_n$ has volume $|\det(T)|$.

So on any finite dimensional vector space Lebesgue measure is uniquely defined up to a positive multiple. In particular, if z is any simplex the linear structure defines Lebesgue measure λ_z on the subsets of z which is unique when normalized by the condition

$$\lambda_z(z) = 1. \tag{5.71}$$

Just use the barycentric coordinate map $c_z : z \to \mathbf{R}^V$, where V is the vertex set of z, and then translate by a point of $c_z(z) = \Delta_V$ to obtain a map from z to $\mathbf{R}_0^V$. As the image has a nonempty interior, we can pull back the Lebesgue measure on $\mathbf{R}_0^V$ and multiply by a positive constant to obtain (5.71). The resulting measure is independent of the choice of translation by translation invariance of Lebesgue measure.

We will call λ_z the Lebesgue measure on the simplex z and refer to it as a k dimensional Lebesgue measure if dim $z = k$.

8 Lemma. *Let $g : K^* \to K$ be a simplicial dynamical system. Let $z^* \in K^*$ and $z_0, z_1 \in K$ with $z_0 \in J(z^*)$ and $z_1 = g(z^*)$. Let λ_{z_0} and λ_{z_1} be the Lebesgue measures on the simplices z_0 and z_1, respectively. If A is a Borel subset of z_1 then*

$$\lambda_{z_1}(A) = 0 \Rightarrow \lambda_{z_0}(g^{-1}(A) \cap z^*) = 0. \tag{5.72}$$

If $\dim z_0 = \dim z_1$ *then for any Borel subset A of z_1*

$$\lambda_{z_0}(g^{-1}(A) \cap z^*) = \lambda_{z_1}(A)\lambda_{z_0}(z^*). \tag{5.73}$$

Proof. If $\dim z^* = \dim z_1$ then $g : z^* \to z_1$ is a linear isomorphism and so by the normalization condition:

$$g_*\lambda_{z^*} = \lambda_{z_1}, \text{ i.e.}$$

$$\lambda_{z^*}(g^{-1}(A) \cap z^*) = \lambda_{z_1}(A). \tag{5.74}$$

If $\dim z^* = \dim z_0 = k$ then both λ_{z^*} and z_{z_0} are k dimensional Lebesgue measures and so on subsets of z^* they differ only by normalization. Hence, for any Borel subset $\tilde{A}$ of z^* we have

$$\lambda_{z^*}(\tilde{A}) = \lambda_{z_0}(\tilde{A})/\lambda_{z_0}(z^*). \tag{5.75}$$

Since $\dim z_0 \geq \dim z^* \geq \dim z_1$ the condition $\dim z_0 = \dim z_1$ implies that both (5.74) and (5.75) hold and so (5.73) follows.

On the other hand, if $\dim z_0 > \dim z^*$ then $\lambda_{z_0}(\tilde{A}) = 0$ for any Borel subset of z^*. In particular, $\lambda_{z_0}(g^{-1}(A) \cap z^*) = 0$ for any Borel subset A of z_1.

Finally, if $k = \dim z_0 = \dim z^* > \dim z_1$ then (5.75) holds but z^* is a degenerate simplex. It suffices by (5.75) to prove that $\lambda_{z^*}(g^{-1}(A) \cap z^*) = 0$ if $\lambda_{z_1}(A) = 0$. Since the boundary ∂z^* has dimension $k-1$ it has measure zero and so we need only consider $\lambda_{z^*}(g^{-1}(A) \cap (z^*)^\circ)$. But by Proposition 2.2, on $(z^*)^\circ$ the linear map g to $(z_1)^\circ$ is diffeomorphic to a projection map. By Fubini's Theorem the preimage of a measure zero subset by projection has measure zero, and a diffeomorphism takes sets of measure zero to sets of measure zero. (5.72) follows. □

Now let B^* be a basic set for G^* with B the associated basic set for G. With $j_{B^*}, g_{B^*} : B^* \to B$ the restricted maps we define the BB^* matrix $L(j_{B^*})$ with support j_{B^*}:

$$L(j_{B^*})_{zz^*} = \begin{cases} \lambda_z(z^*) & j(z^*) = z \\ 0 & \text{otherwise,} \end{cases} \tag{5.76}$$

i.e. in general, $L(j_{B^*})_{zz^*} = \lambda_z(z^* \cap z)$. Define

$$L(G^*_{B^*}) = L(j_{B^*})^T K(g_{B^*})$$

$$L(G_B) = K(g_{B^*}) L(j_{B^*})^T \tag{5.77}$$

with supports $G^*_{B^*}$ and G_B respectively.

As in (5.55) let $\tilde{l}$ and $\tilde{r}$ be the left and right positive eigenvectors for $L(G_B)$ with positive eigenvalue $\tilde{\gamma}$, i.e.

$$\tilde{l}^T L(G_B) = \tilde{l}^T K(g_{B^*}) L(j_{B^*})^T = \tilde{\gamma} \tilde{l}^T$$

$$L(G_B)\tilde{r} = K(g_{B^*}) L(j_{B^*})^T \tilde{r} = \tilde{\gamma}\tilde{r}. \tag{5.78}$$

With $u_z = 1$ for all $z \in B$ we can normalize so that

$$\tilde{l}^T \tilde{r} = 1 = u^T \tilde{r}. \tag{5.79}$$

Now define $\tilde{l}^*$ and $\tilde{r}^*$ vectors on B^* by:

$$\tilde{l}^{*T} = l^T K(g_{B^*}) \text{ and } \tilde{r}^* = \frac{1}{\tilde{\gamma}} L(j_{B^*})^T \tilde{r}.$$

$$\tilde{l}^*_{z^*} = \tilde{l}_{g(z^*)} \text{ and } \tilde{r}^*_{z^*} = \frac{1}{\tilde{\gamma}} \lambda_{j(z^*)}(z^*) \tilde{r}_{j(z^*)} \tag{5.80}$$

for all $z^* \in B^*$. As in (5.58), $\tilde{l}^*$ and $\tilde{r}^*$ are positive left and right eigenvalues for $L(G^*_{B^*})$ with eigenvalue $\tilde{\gamma}$. As in (5.61) we have $\tilde{l}^{*T}\tilde{r}^* = 1$.

9 Proposition. *With B^* a G^* basic set and B the associated G basic set let $g_{|B^*|} : |B^*| \to |B|$ denote the restriction of the p.l map g on X to $|B^*| = \cup\{z^* \in B^*\}$. For A any Borel subset of $|B|$ let $\lambda_\#(A)$ denote the B vector whose z component is $\lambda_z(A \cap z)$ for all $z \in B$.*

(5.81) $$\lambda_\#(A)^T L(G_B) = \lambda_\#((g_{|B^*|})^{-1}(A))^T.$$

Proof. For $z \in B$ the z component of the product vector on the left side of (5.81) is

$$\sum_{z_1 \in B, z^* \in B^*} \lambda_{z_1}(A \cap z_1) K(g_{B^*})_{z_1 z^*} \lambda_z(z^* \cap z)$$

(5.82) $$= \sum_{z^* \in B^*} \lambda_z(z^* \cap z) \lambda_{g(z^*)}(A \cap g(z^*)).$$

As in (5.76), $\lambda_z(z^* \cap z) = 0$ unless $j(z^*) = z$ so this sum equals

$$\sum\{\lambda_z(z^*)\lambda_{g(z^*)}(A \cap g(z^*)) : j(z^*) = z\}$$

(5.83) $$= \sum\{\lambda_z(g^{-1}(A) \cap z^*) : j(z^*) = z\}$$

by (5.73). For $z \in B$, $|B^*| \cap (z)^\circ = (\cup\{z^* \in B^* : j(z^*) = z\}) \setminus \partial z$. For distinct B^* simplices z_1^* and z_2^* the intersection $z_1^* \cap z_2^* \cap z$ need not be empty but has dimension smaller than that of z and so has λ_z measure zero. Consequently, the latter sum in (5.83) is

(5.84) $$\lambda_z(g^{-1}(A) \cap |B^*|) = \lambda_z((g_{|B^*|})^{-1}(A)),$$

proving (5.81). □

We apply these results first to the case where B^* is a polyhedral basic set. Recall from Corollary 4.25 that B^* is called polyhedral when $|B^*| = |B|$. So for every $z \in B$:

(5.85) $$z = \cup\{z^* \in B^* : j(z^*) = z\}$$

because the union contains $(z)^\circ \cap |B^*|$ in any case. Since distinct B^* simplices intersect in a subset of the boundaries it follows that

$$(5.86) \qquad 1 = \lambda_z(z) = \sum \lambda_z(z^*) = \sum L(j_{B^*})_{zz^*}$$

summing over all $z^* \in B^*$ or, equivalently, over all $z^* \in B^*$ such that $j(z^*) = z$. Thus, for a polyhedral basic set B^* the matrices $L(G^*)$ and $L(G)$ of (5.77) are stochastic and so u^T is a left eigenvector with eigenvalue 1. Hence by (5.79) and uniqueness in Perron-Frobenius Theorem:

$$(5.87) \qquad \tilde{\gamma} = 1 \text{ and } \tilde{l} = u$$

when B^* is polyhedral.

The associated distribution vectors $\tilde{p}$ on B and $\tilde{p}^*$ on $B*$ are given by:

$$(5.88) \qquad \tilde{p}_z = \tilde{r}_z \text{ and } \tilde{p}^*_{z^*} = \tilde{r}^*_{z^*} = \lambda_{j(z^*)}(z^*)\tilde{p}_{j(z^*)}.$$

On $|B| = |B^*|$ we define the invariant Lebesgue measure λ_B by

$$(5.89) \qquad \lambda_B(A) = \sum_{z \in B} \tilde{p}_z \lambda_z(A \cap z)$$

for every Borel subset A of $|B|$.

10 Theorem. *For a simplicial dynamical system $g : K^* \to K$ let B^* be a polyhedral basic set for G^*, i.e. $|B^*| = |B|$ where B is the associated basic set for G. Let ν^L be the $s^+_{G^*}$ invariant Markov measure on $B^{*+}_{G^*}$ associated with the stochastic matrix $L(G^*) = L(j_{B^*})^T K(g_{B^*})$, and let λ_B be the invariant Lebesgue measure on $|B|$.*

$$(5.90) \qquad h^+_* \nu^L = \lambda_B.$$

In particular, $h^+ : (B^{*+}_{G^*}, s^+_{G^*}, \nu^L) \to (|B|, g, \lambda_B)$ is an isomorphism of ergodic measurable dynamical systems.

Proof. For $\tilde{q}$ any distribution vector on the set B we can define the measure $\lambda_{\tilde{q}}$ on $|B|$ by

$$(5.91) \qquad \lambda_{\tilde{q}}(A) = \lambda_\#(A)^T \tilde{q}$$

for every Borel subset A of $|B|$. By (5.81)

$$(5.92) \qquad \lambda_{\tilde{q}}((g_{|B^*|})^{-1}(A)) = \lambda_\#(A)^T L(G_B)\tilde{q}.$$

Because $\tilde{p}$ is a right eigenvector with eigenvalue $\gamma = 1$, (5.92) implies that $\lambda_{\tilde{p}} = \lambda_B$ is $g_{|B^*|}$ invariant.

Conversely, if $\lambda_{\tilde{q}}$ is invariant then with $A = z \in B$ (5.92) implies

$$\lambda_{\tilde{q}}(z) = \lambda_{\tilde{q}}((g_{|B^*|})^{-1}(z)) = \lambda_{\#}(z)^T L(G_B)\tilde{q}. \tag{5.93}$$

But $\lambda_{\#}(z)$ is the basis vector e^z in $\mathbf{R}^B$. Hence, $\lambda_{\tilde{q}}(z) = \tilde{q}_z$ and so (5.93) implies that $\tilde{q}$ is a right eigenvector with eigenvalue 1, i.e. $\tilde{q} = \tilde{p}$.

To prove (5.90) we select a G^* chain $\{z_0^*, \dots, z_n^*\}$ in B^*. We compare the cylinder set C in $B_{G^*}^{*+}$ and the simplex Δ in $|B|$ where

$$C = (\pi_0^+)^{-1}(z_0^*) \cap \dots \cap (\pi_n^+)^{-1}(z_n^*)$$

$$\Delta = \cap_{i=0}^n g^{-i}(z_i^*) = \overline{g}_{z_0^*} \circ \dots \circ \overline{g}_{z_{n-1}^*}(z_n^*). \tag{5.94}$$

Let $\mathbf{z}^* \in B_{G^*}^{*+}$ and $x = h^+(\mathbf{z}^*)$. Clearly, $\mathbf{z}^* \in C$ implies $x \in \Delta$. On the other hand, if B^* is a k skeleton basic set and $\mathbf{z}^* \in \mathrm{Int}_k^+$ then by Proposition 4.10 $\mathbf{z}^* = q^{*+}(x)$. So if $x \in \Delta$ then by (2.18) $q^*(g^i(x)) = z_i^*$ $(i = 0, \dots, n)$ and so $\mathbf{z}^* \in C$. It follows that

$$(h^+|B_{G^*}^{*+})^{-1}(\Delta) \cap \mathrm{Int}_k^+ \subset C \subset (h^+)^{-1}(\Delta). \tag{5.95}$$

By Theorem 2c, $\nu^L(\mathrm{Int}_k^+) = 1$ and so

$$\nu^L((h^+|B_{G^*}^{*+})^{-1}(\Delta)) = \nu^L(C). \tag{5.96}$$

So by (5.45)

$$\begin{aligned} &\nu^L((h^+|B_{G^*}^{*+})^{-1}(\Delta)) = \\ &L(G_{B^*}^*)_{z_n^* z_{n-1}^*} \dots L(G_{B^*}^*)_{z_1^* z_0^*} \tilde{p}_{z_0^*}^* \\ &\tilde{p}_{z_0} \lambda_{z_0}(z_0^*) \lambda_{z_1}(z_1^*) \dots \lambda_{z_n}(z_n^*) \end{aligned} \tag{5.97}$$

where $z_i = j(z_i^*)$ for $i = 0, \dots, n$ and so $z_i = g(z_{i-1}^*)$ for $i = 1, \dots, n$. We use (5.88) for $\tilde{p}_{z_0^*}^*$ and (5.76), (5.77) for $L(G_{B^*}^*)_{z_i^* z_{i-1}^*}$ $(i = 1, \dots, n)$. On the other hand, by (5.73) and induction

$$\lambda_{z_0}(\Delta) = \lambda_{z_0}(z_0^*) \lambda_{z_1}(z_1^*) \dots \lambda_{z_n}(z_n^*). \tag{5.98}$$

So from (5.89) we have

$$\lambda_B(\Delta) = \nu^L((h^+|B_{G^*}^{*+})^{-1}(\Delta)). \tag{5.99}$$

Since these cylinders generate the Borel subsets of $B^{*+}_{G^*}$, equation (5.90) follows.

The Markov measure ν^L is ergodic by Lemma 5. Ergodicity is a measure isomorphism invariant and so λ_B is ergodic.

Remark. The computation in the beginning of the proof, shows that the right eigenvector $\tilde{r} = \tilde{p}$ is the unique distribution vector $\tilde{q}$ on B such that $\lambda(A) = \sum_{z \in B} \tilde{q}_z \lambda_z(A \cap z)$ defines a g invariant measure. This is why we speak of λ_B as *the* invariant Lebesgue measure on $|B|$. □

We pause to consider the effects of replacing the subdivision K^* of K by an isomorphic subdivision K_1^* of K. If $r_1 : K^* \to K_1^*$ is the subdivision isomorphism and $g_1 : K_1^* \to K$ is the simplicial dynamical system with $g_1 \circ r_1 = g$ then in Theorem 3.10 there is constructed a homeomorphism ρ_1 on X which maps g on X to g_1 on X, i.e. $\rho_1 \circ g = g_1 \circ \rho_1$. The continuous map ρ_1 satisfies, and so is characterized by, the commutative diagram (3.91).

11 Proposition. *Let $K_1^* \in \mathrm{Iso}(K^* : K)$ and $g_1 : K_1^* \to K$ satisfy $g_1 \circ r_1 = g$ where $r_1 : K^* \to K_1^*$ is the subdivision isomorphism. Let ρ_1 be the conjugating homeomorphism on X described by Theorem 2.10.*

If B^* is a k skeleton basic set for G^* with B the associated basic set for G then $B_1^* = r_1(B^*) \subset K_1^*$ is the k skeleton basic set for G_1^* whose associated G basic set is B.

$$\rho_1(|B^*|) = |B_1^*|$$

$$\rho_1(h^+(B^{*+}_{G^*})) = h_1^+(B^{*+}_{1G_1^*}). \tag{5.100}$$

If B^* is an interior basic set for G^* and μ_{B^*} is the Parry measure on $h^+(B^{*+}_{G^*})$ then B_1^* is an interior basic set for G_1^* and

$$\rho_{1*}\mu_{B^*} = \mu_{B_1^*} \tag{5.101}$$

where the latter is the Parry measure on $h_1^+(B^{*+}_{G_1^*})$.

If B^* is a polyhedral basic set for G^* then B_1^* is polyhedral for G_1^*.

Proof. From (3.89) it is clear that r_1 maps G^* to G_1^* and so B_1^* is a G_1^* basic set. Since $r_1 \times 1_K$ maps g and j to g_1 and j_1 it follows that B is the associated G basic set in each case. From (3.78) we have

$$\rho_1(z^*) = r_1(z^*) \tag{5.102}$$

i.e. the image of the subset z^* under the continuous map ρ_1 is the simplex $r_1(z^*) \in K_1^*$. The first equation of (5.100) follows and the second follows from the commutative diagram (3.91). It then follows that B_1^* is interior (or polyhedral) if B^* is.

Because the finite set bijection $r_1 : B^* \to B_1^*$ maps the relation $G_{B^*}^*$ to $G_{1B_1^*}^*$ it follows that $r_1^+ : B_{G^*}^{*+} \to B_{1G_1^*}^{*+}$ induces an isomorphism of subshifts of finite type. Up to the name change, the matrices $K(G^*)$ and $K(G_1^*)$ are the same and so the Markov measures ν^{P^*} and $\nu^{P_1^*}$ are related by $(r_1^+)_*$. Then (5.101) follows from the commutative diagram (3.91) and the definition in Theorem 6 of the Parry measures μ_{B^*} and $\mu_{B_1^*}$. □

In the polyhedral case it is usually *not* true that the restriction of ρ_1 to $|B^*| = |B| = |B_1^*|$ maps the Lebesgue measure λ_{B^*} to $\lambda_{B_1^*}$. In fact, the $\rho_{1*}\lambda_{B^*}$ and the Lebesgue measure $\lambda_{B_1^*}$ are usually mutually singular. This is a generalization of an odd phenomenon discussed in Milnor (1997) and first observed by Katok.

The problem is that the Markov measures $\nu^{\tilde{P}^*}$ from $L(G^*)$ on $B_{G^*}^{*+}$ and $\nu^{\tilde{P}_1^*}$ from $L(G_1^*)$ on $B_{1G_1^*}^{*+}$ are usually associated with different matrices. Any two distinct ergodic measures are mutually singular and these measures often have different entropy and so are not even isomorphic much less equal.

Consider the *tent map*. This is a variation, considered by Yorke, of the case discussed by Milnor. Here $K = [0,1]$ and K^* is the barycentric subdivision with new vertex $\frac{1}{2}$. $g : K^* \to K$ is defined by $g(0) = g(1) = 0$ and $g(\frac{1}{2}) = 1$. $B^* = \{[0,\frac{1}{2}], [\frac{1}{2},1]\}$ has associated basic set $B = \{[0,1]\}$. The shift $s_{G^*}^+$ on $B_{G^*}^{*+}$ is the full shift on two symbols. By (5.89) the Lebesgue measure λ_B is λ_z with $z = [0,1]$, ordinary Lebesgue measure on $[0,1]$. The matrix $L(G_{B^*}^*)$ is the stochastic retraction P of $K(G_{B^*}^*)$. The associated Markov measure $\nu^P = \nu^{\tilde{P}}$ on $B_{G^*}^{*+}$ is the so-called *Bernoulli measure* with probabilities $\{\frac{1}{2},\frac{1}{2}\}$ and $\lambda_B = \mu_{B^*}$ is the Parry measure.

Suppose K_1^* is the derived subdivision with new vertex at p $(0 < p < \frac{1}{2})$ instead of $\frac{1}{2}$. Thus $r_1 : K^* \to K_1^*$ is the isomorphism with $r_1(0) = 0$, $r_1(\frac{1}{2}) = p$ and $r_1(1) = 1$. $B_1^* = \{[0,p],[p,1]\}$. The shift $s_{G^*}^+$ on $B_{1G_1^*}^+$ is the full shift on two symbols and r_1^+ is just the isomorphism between the two shifts obtained by replacing the alphabet B^* by B_1^*. So the Parry measure $\nu^{P_1} = r_{1*}\nu^P$ and by (5.101)

$$\rho_{1*}\lambda_B = \rho_{1*}\mu_{B^*} = \mu_{B_1^*}. \tag{5.103}$$

Now (5.89) applies to g_1 and so $\lambda_{B_1} = \lambda_z$ is still ordinary Lebesgue measure on $[0,1]$. But the matrix $L(G^*_{1B^*_1})$ yields the Bernoulli measure $\nu^{\tilde{P}_1}$ with probabilities $\{p, 1-p\}$ and $\lambda_{B_1} = (h^+_{B^*_1})_* \nu^{\tilde{P}_1}$. To summarize:

(1) ρ_1 is a homeomorphism on $[0,1]$ with $\rho_1 \circ g = g_1 \circ \rho_1$ and so the topological dynamical systems given by g and g_1 on $[0,1]$ are conjugate by ρ_1.

(2) Lebesgue measure λ on $[0,1]$ is an invariant measure for g and for g_1.

(3) It is not true that $\rho_{1*}\lambda = \lambda$. In fact, the measure dynamical systems $([0,1], g, \lambda)$ and $([0,1], g_1, \lambda)$ are not even isomorphic. There is a measure isomorphism of the first to a Bernoulli shift with entropy ln 2 and of the second to a Bernoulli shift with entropy $-p \ln p - (1-p)\ln(1-p)$.

(4) It follows that ρ_1 is neither differentiable nor p.l.

In this case among the isomorphisms of g there exists one with respect to which λ_B is the Parry measure on $|B|$ and so λ_B is an invariant measure for g_{B^*} on $h^+(B^{*+}_{G^*}) = |B|$ with maximal entropy. We will later see examples where the maximal entropy cannot be so achieved using a Lebesgue measure.

Return now to the general case. Suppose that B^* is a k skeleton basic set for G^* with associated G basic set B, but assume that B^* is *tattered*, i.e. $|B^*|$ is a proper subset of B. Let $\tilde{l}$, $\tilde{r}$ and $\tilde{\gamma}$ be the eigenvectors and eigenvalue for $L(G_B)$ described in (5.78) and (5.79). We define the Lebesgue measure λ_B on $|B|$: For A a Borel subset of $|B|$, define

$$\lambda_B(A) = \sum_{z \in B} \tilde{r}_z \lambda_z(A \cap z)$$

(5.104)
$$= \lambda_{\#}(A)^T \tilde{r}$$

where the latter equation uses the notation of Proposition 9. Observe that in the polyhedral case, $\tilde{l}_z = 1$ and $\tilde{p}_z = \tilde{r}_z$ so that (5.104) extends (5.89). By (5.92)

(5.105)
$$\lambda_B((g_{|B^*|})^{-1}(A)) = \tilde{\gamma}\lambda_B(A).$$

Because $u^T\tilde{r} = 1$, we have

(5.106)
$$\lambda_B(|B|) = 1.$$

Applying (5.105) with $A = |B|$ so that $(g_{|B^*|})^{-1}(A) = |B^*|$ we see that

$$\lambda_B(|B^*|) = \tilde{\gamma}. \tag{5.107}$$

Because B^* is tattered it follows that the eigenvalue is strictly less than 1. Let $z^* \in K^*$ with $\dim z^* = k$ and $j(z^*) = z \in B$ but $z^* \notin B^*$. Since $|B^*| \cap (z^*)^\circ = \emptyset$ and $\lambda_z(z^*) = \lambda_z((z^*)^\circ)$,

$$\lambda_B(|B^*|) \leq 1 - \tilde{r}_z \lambda_z(z^*) < 1. \tag{5.108}$$

Thus, in the tattered case λ_B is not an invariant measure for g.

12 Proposition. *Let $g : K^* \to K$ be a simplicial dynamical system, B^* be a basic set for G^* with B the associated basic set for G. Let λ_B be the natural Lebesgue measure on $|B|$.*

If B^* is tattered then the basic set image $h^+(B^{*+}_{G^*})$ is a closed nowhere dense subset of $|B|$ and

$$\lambda_B(h^+(B^{*+}_{G^*})) = 0. \tag{5.109}$$

Proof. Let $F_0 = |B|$ and define, inductively, for $n - 1, 2, \ldots$

$$F_n = |B^*| \cap g^{-1}(F_{n-1}) = (g_{|B^*|})^{-1}(F_{n-1}). \tag{5.110}$$

By (5.106) $\lambda_B(F_0) = 1$. Inductively, (5.105) implies

$$\lambda_B(F_n) = \tilde{\gamma}\lambda_B(F_{n-1}) = \tilde{\gamma}^n. \tag{5.111}$$

By (5.107) and (5.108), $\tilde{\gamma} < 1$ and so the closed subset $h^+(B^{*+}_{G^*})$ of $\cap_{n=0}^{\infty} F_n$ has λ_B measure zero.

Since any open subset of $|B|$ has positive λ_B measure it follows from (5.109) that $h^+(B^{*+}_{G^*})$ is nowhere dense. □

It follows that if B^* is a tattered k skeleton basic set then for $z \in K$

$$\dim z \geq k \Rightarrow \lambda_z(h^+(B^{*+}_{G^*}) \cap z) = 0. \tag{5.112}$$

This follows immediately from (5.109) for $z \in B$. If $z \notin B$ then dim $z \geq k$ implies $z \cap |B| \subset \partial z$ and so has λ_z measure 0.

In general we say that a Borel subset A of X has *Lebesgue measure zero with respect to* K if

$$\lambda_z(A \cap z) = 0 \text{ for all } z \in K. \tag{5.113}$$

This notion depends on the triangulation K. For example, such a subset must be disjoint from the set of vertices of K.

13 Lemma. *Let $g : K^* \to K$ be a simplicial dynamical system. If A has Lebesgue measure zero with respect to K then $g^{-1}(A)$ does also.*

Proof. If $z \in K$ with dim $z = k$ then

$$\lambda_z(g^{-1}(A)) = \sum \lambda_z(g^{-1}(A) \cap z^*) \tag{5.114}$$

summing over $z^* \in K^*$ with dim $z^* = k$ and $j(z^*) = z$. By (5.72) each term of the sum is zero when A has Lebesgue measure zero with respect to K. □

14 Theorem. *Let $g : K^* \to K$ be a simplicial dynamical system. For $x \in X = |K| = |K^*|$ the endset $B^*(x)$ is the G^* basic set such that $q^*(g^i(x)) \in B^*(x)$ for i sufficiently large. $\{x : B^*(x)$ is tattered$\}$ is a Borel set of first category and of Lebesgue measure zero with respect to K. There is a dense G_δ subset D of X with $\lambda_z(D \cap z) = 1$ for every $z \in K$ and such that $x \in D$ implies the endset of x is a polyhedral basic set for G^*.*

Proof. Let B^* be a k skeleton basic set for G^* with associated G basic set B. Let $F(B^*) = h^+(B^*_{G^*}) \setminus |\overline{S}^{k-1}(K)|$. As the intersection of a closed set with an open set, $F(B^*)$ is an F_σ and a G_δ subset. By Proposition 4.13b, $q^{*+}(x) \in B^{*+}_{G^*}$ iff $x \in \cap_{i=0}^{\infty} g^{-i}(F(B^*))$. Hence,

$$B^*(x) = B^* \Leftrightarrow x \in \cup_{n=0}^{\infty} \cap_{i=n}^{\infty} g^{-i}(F(B^*)) \tag{5.115}$$

$$\{x : B^*(x) = B^*\} \subset \cup_{n=0}^{\infty} g^{-n}(F(B^*)). \tag{5.116}$$

Taking the union over the finitely many tattered sets we see that $\{x : B^*(x)$ is tattered$\}$ is a Borel set and it is contained in the F_σ set:

$$\cup_{B^* \text{ tattered}} \cup_{n=0}^{\infty} g^{-n}(F(B^*)). \tag{5.117}$$

If B^* is a k skeleton basic set and dim $z < k$ then $z \cap (h^+(B^{*+}_{G^*}) \setminus |\overline{S}^{k-1}(K)|) =$

$\emptyset$. So if B^* is tattered, (5.112) implies that $F(B^*)$ has Lebesgue measure 0 with respect to K. By Lemma 13 and induction each $g^{-n}(F(B^*))$ has Lebesgue measure 0 with respect to K. So the countable union in (5.117) is an F_σ subset of Lebesgue measure 0. A fortiori it is a countable union of closed nowhere dense subsets. The complement of this F_σ set is the required dense G_δ subset D. □

A 0 skeleton basic set is a periodic orbit of vertices of K, and since $B^* = B$ it is clearly polyhedral. Corollary 4.25 implies that when g is nondegenerate every k skeleton basic set terminal among k skeleton basic sets is polyhedral. In the degenerate case it can happen that only the 0 skeleton basic sets are polyhedral.

15 Corollary. *Let $g : K^* \to K$ be a simplicial dynamical system such that every k skeleton basic set with $k > 0$ is tattered. There exists a dense G_δ subset D of $X = |K| = |K^*|$ with $\lambda_z(D \cap z) = 1$ for every $z \in K$ and such that $x \in D$ implies $g^i(x)$ is a vertex of K for some $i \in \mathbf{Z}_+$.*

Proof. $B^*(x)$ is a 0 skeleton basic set iff $g^i(x)$ is a vertex for some $i \in \mathbf{Z}_+$. □

6. Generalized Simplicial Dynamical Systems

Let $g : K^* \to K$ be a simplicial map with $|K| = X$ and K^* a subdivision of K. We have defined g to be a simplicial dynamical system if K^* is a proper subdivision. As we saw in Theorem 3.2 the closed relation $\hat{h}^+ : \hat{K}^+_{\hat{G}} \to X$ is then a function. As we will see it is this latter condition which is really required for all our results. The proper subdivision condition is easy to check and suffices for almost all applications. The more general situation does turn up in two cases.

First, suppose not that every simplex z^* of K^* is properly included in K (see Definition 2.7) but only that the nondegenerate simplices are properly included or even that only the G^* periodic simplices, the elements of $|\mathcal{O}G^*|$, are properly included in K. Observe that this condition depends on g as well as on K^* and K. For any $\mathbf{z}^* \in K^{*+}_{G^*}$ if a simplex $w^* \in K^*$ occurs twice in the sequence $\{z_i^*\}$ then it is G^* periodic. So the number of simplices in the sequence which are not properly included in K is finite. In fact it is bounded by the cardinality of K^*. It is easy to adapt the proof of Theorem 3.2 to show that for $(\mathbf{z}^*, \mathbf{t}) \in \hat{K}^+_{\hat{G}}$ there exists at most one point $x \in X$ such that $g^i(x) \in \langle z_i^*, t_i \rangle$ for all $i \in \mathbf{Z}_+$.

For the second case, suppose that $g : K^* \to K$ is nondegenerate. If $z^* \in K^*$ then $z = g(z^*)$ is subdivided by K^* into the subcomplex $J^{-1}(z)$ of K^*. We can use the linear isomorphism $g : z^* \to z$ to pull this triangulation back to z^*. We thus obtain a subdivision K^{**} of K^* such that $g : K^{**} \to K^*$ is a simplicial map and so $g^2 : K^{**} \to K$ is simplicial. As it is nondegenerate we can repeat the procedure pulling K^* back via g^2 to obtain K^{***} a subdivision of K^{**} such that $g : K^{***} \to K^{**}$ is simplicial and so $g^3 : K^{***} \to K$ is simplicial. We can, of course, continue inductively.

It sometimes happens that while K^* is not a proper subdivision of K, either K^{**}, K^{***} or one of these higher refinements is a proper subdivision of K. If K^{**} is a proper subdivision of K then Theorem 3.2 applies directly to $g^2 : K^{**} \to K$ and from this it easily follows that for $(\mathbf{z}^*, \mathbf{t}) \in \hat{K}^+_{\hat{G}}$ there is a unique $x \in X$ such that $g^i(x) \in \langle z_i^*, t_i \rangle$ for all $i \in \mathbf{Z}_+$.

In the latter case it need not be true that each $\overline{g}_{z^*,t}$ is a contraction but by iterating in pairs or triplets we obtain contractions. This turns out to be the general case.

1 Theorem. *Let $g : K^* \to K$ be a simplicial map with K^* a subdivision* *of* K. *Let* $\hat{h}$ *and* $\hat{h}^+$ *be the sample path relations defined by (3.15). The following conditions are equivalent. When they hold we call* g *a* generalized simplicial dynamical system.

(1) $\hat{h}$ *is a continuous, surjective map from* $\hat{K}_{\hat{G}}$ *to* X_g.

(2) $\hat{h}^+$ *is a continuous, surjective map from* $\hat{K}^+_{\hat{G}}$ *to* X.

(3) There exists a positive integer m *and a positive real number* $\lambda < 1$ *such that for* $\{(z_0^*, t_0), \ldots, (z_m^*, t_m)\}$ *a* $\hat{G}$ *chain in* $\hat{K}$

$$
\begin{aligned}
& d_K(\overline{g}_{z_0^*,t_0} \circ \ldots \circ \overline{g}_{z_m^*,t_m}(y_1), \overline{g}_{z_0^*,t_0} \circ \ldots \circ \overline{g}_{z_m^*,t_m}(y_2)) \\
(6.1) \qquad & \leq \lambda d_K(y_1, y_2) \text{ for all } y_1, y_2 \in g(z_m^*).
\end{aligned}
$$

Proof. By Theorem 3.2c, the relations $\hat{h}$ and $\hat{h}^+$ are closed and surjective. So if one is a function then it is a continuous surjection. Adapting the proof of Theorem 3.2d, it is easy to see that (3) $\Rightarrow$ (1) and (2).

Consider the set of triples $DK^* \subset K^* \times K^* \times K^*$ defined by (4.26). We define a similar subset

$$
\begin{aligned}
& PK^* = \{(z_1^*, z_2^*, z_3^*) \in K^* \times K^* \times K^* : \\
(6.2) \qquad & z_2^* \text{ is nondegenerate}, z_1^* < z_2^*, z_3^* < z_2^* \text{ and } z_1^* \cap z_3^* = \emptyset\}.
\end{aligned}
$$

On the finite set PK^* define the relation

$$
(6.3) \qquad PG^* = (G^* \times G^* \times G^*) \cap (PK^* \times PK^*).
$$

Suppose $(w_1^*, w_2^*, w_3^*) \in PG^*(z_1^*, z_2^*, z_3^*)$. Set $(z_1, z_2, z_3) = (g(z_1^*), g(z_2^*), g(z_3^*))$ $K \times K \times K$. Since z_2^* is nondegenerate g is a linear isomorphism on z_2^* and so z_1 and z_3 are disjoint faces of z_2. Furthermore, $w_1^* \subset z_1 \cap w_2^*$ and $w_3^* \subset z_3 \cap w_2^*$. This means that w_2^* is *not* properly included in K. Thus, if K^* is a proper subdivision of K then the relation PG^* on PK^* is the empty relation. In general, a closed relation on a compact space is called *nilpotent* if some iterate is empty (see Akin (1993) p. 37). We now show that the three conditions described in the statement of the theorem occur exactly when PG^* is a nilpotent relation.

Fix m larger than the cardinality of the set PK^*. By the pigeonhole principle any PG^* chain of at least m terms must contain a repeat. So either there exists no PG^* chain of length m, the nilpotent case, or there exist periodic chains for PG^*.

Case 1. There exists a PG^* chain $\{(z_{10}^*, z_{20}^*, z_{30}^*), \dots, (z_{1k}^*, z_{2k}^*, z_{3k}^*)\}$ in PK^* with $(z_{10}^*, z_{20}^*, z_{30}^*) = (z_{1k}^*, z_{2k}^*, z_{3k}^*)$.

Extend these periodically to get three elements $\mathbf{z_1^*}, \mathbf{z_2^*}, \mathbf{z_3^*} \in K_{G^*}^*$ such that for all i, $(z_{1i}^*, z_{2i}^*, z_{3i}^*) \in PK^*$. Let $\xi \in h(\mathbf{z_1^*}) = \hat{h} \circ \hat{p}^{-1}(\mathbf{z_1^*})$ and $\eta \in h(\mathbf{z_3^*})$. For every $i \in \mathbf{Z}$

$$\xi_i \in z_{1i}^* \subset z_{2i}^*$$

$$\eta_i \in z_{3i}^* \subset z_{2i}^* \tag{6.4}$$

By Theorem 3a,c h is a surjective relation and so some such ξ and η exist. Since z_{1i}^* and z_{3i}^* are disjoint faces of z_{2i}^* we have for all i

$$\xi_i \neq \eta_i. \tag{6.5}$$

Thus, ξ, $\eta \in h(\mathbf{z_2^*})$. Because z_{2i}^* is nondegenerate for all i, there is a unique $(\mathbf{z}^*, \mathbf{t}) \in \hat{K}_{\hat{G}}$ which projects to $\mathbf{z_2^*}$. Hence, ξ and η are distinct elements in $\hat{h}(\mathbf{z}^*, \mathbf{t})$. ξ_0 and η_0 are distinct elements in $\hat{h}^+(\pi_{\hat{G}}^+(\mathbf{z}^*, \mathbf{t}))$. Hence, neither (1) nor (2) holds in Case 1. Because (3) $\Rightarrow$ (1) and (2) it follows that (3) fails as well.

Case 2. There is a positive integer m such that $(PG^*)^m = \emptyset$, i.e. there is no PG^* chain of length m.

We will show that in this case if $\{z_0^*, \dots, z_m^*\}$ is a G^* chain consisting of nondegenerate simplices then there exists $\lambda < 1$, depending on the chain, such that

$$\begin{aligned} &d_K(\overline{g}_{z_0^*} \circ \dots \circ \overline{g}_{z_m^*}(y_1), \overline{g}_{z_0^*} \circ \dots \circ \overline{g}_{z_m^*}(y_2)) \\ &\leq \lambda d_K(y_1, y_2). \end{aligned} \tag{6.6}$$

Since there are only finitely many such chains (m is fixed here), we can use the largest λ to get the contraction rate independent of the choice of chain.

To get (6.1) from (6.6) we consider for each $i = 0, \dots, m$ elements w_i^* nondegenerate faces of z_i^* such that $g(w_i^*) = g(z_i^*)$. Clearly, $\{w_0^*, \dots, w_m^*\}$ is a G^* chain of nondegenerate simplices to which (6.6) applies. By Corollary

2.4 the linear map $\overline{g}_{z_0^*,t_0} \circ \ldots \circ \overline{g}_{z_m^*,t_m}$ is a convex combination of the associated composites $\overline{g}_{w_0^*} \circ \ldots \circ \overline{g}_{w_m^*}$. So (6.1) follows from (6.6).

To prove (6.6) we follow the proof of (2.64) in Theorem 2.11. When we construct the stochastic matrix P_{ij} associated with $\overline{g}_{z_0^*} \circ \ldots \circ \overline{g}_{z_m^*}$ it follows that condition (2.46) holds provided that for every distinct pair of vertices $v_{j_1}^*$, $v_{j_2}^*$ of z_m^* there is some vertex v_{i_0} of $j(z_0^*)$ for whom the barycentric coordinate of y_p is positive where $y_p = \overline{g}_{z_0^*} \circ \ldots \circ \overline{g}_{z_{m-1}^*}(v_{j_p}^*)$ $(p = 1, 2)$. It is sufficient to show that the two faces $w_p^* = q^*(y_p)$ $(p = 1, 2)$ of z_0^* intersect for we can then choose v_{i_0} to be any vertex of $j(w_1^* \cap w_2^*)$ in $j(z_0^*)$. Assume instead that $w_1^* \cap w_2^* = \emptyset$. Define for $i = 0, \ldots, m$

$$
\begin{aligned}
z_{1i}^* &= q^*(g^i(y_1)) \\
z_{2i}^* &= z_i^* \\
z_{3i}^* &= q^*(g^i(y_2)).
\end{aligned}
\tag{6.7}
$$

Since $g^i(y_p) \in z_i^*$ for $i = 0, \ldots, m$ and $p = 1, 2$ we have $z_{1i}^* < z_{2i}^*$ and $z_{3i}^* < z_{2i}^*$. By assumption z_{2i}^* is nondegenerate. By Lemma 3.1 $\{z_{10}^*, \ldots, z_{1m}^*\}$ and $\{z_{30}^*, \ldots, z_{3m}^*\}$ are G^* chains as is $\{z_{20}^*, \ldots, z_{2m}^*\}$. By assumption $z_{10}^* \cap z_{30}^* = \emptyset$.

Proceeding inductively if $z_{1i}^* \cap z_{3i}^* = \emptyset$ then $g(z_{1i}^*)$ and $g(z_{3i}^*)$ are disjoint faces of $g(z_{2i}^*)$. As these contain $z_{1(i+1)}^*$ and $z_{3(i+1)}^*$ respectively it follows that $z_{1(i+1)}^* \cap z_{3(i+1)}^* = \emptyset$. Thus we have constructed a PG^* chain in PK^* of length m. This contradicts the assumption of Case 2. It follows that $w_1^* \cap w_2^* \neq \emptyset$ as required.

Thus, in Case 2 condition (3) holds. Since (3) $\Rightarrow$ (1) and (2) the latter hold as well. □

It easily follows that all of the results of the previous chapters hold for generalized simplicial dynamical systems. Only the explicit estimates like (3.34), (3.47) and (3.80) have to be adjusted.

7. Examples

In the examples which follow K consists of a single simplex σ together with all of its faces. The simplicial map $g : K^* \to K$ is surjective in each case and so there is at least one nondegenerate d dimensional simplex in K^* where $d = \dim \sigma$. Hence, $B = \{\sigma\}$ is the unique d dimensional basic set for G. The associated basic set B^* consists of all of the nondegenerate d dimensional simplices of K^*. In each of the figures the simplex σ is two dimensional. The left and right base vertices of the triangle σ are denoted 0 and 1, respectively, and the top vertex is denoted 2. In the figures each vertex of the subdivision K^* (including the vertices of K) is labeled by its image under g.

Example (1) (The Generalized Tent Map). Let X be the d dimensional simplex σ with vertices denoted $0, \ldots, d$. Let K consist of σ together with its faces and let K^* be the first barycentric subdivision of K. Define $g : K^* \to K$ by $g(b(z)) = k$ when $\dim z = k$. So all the vertices of K are mapped to 0, the barycenters of 1 simplices of K, e.g. $b([0, 1])$ are mapped to 1, etc. When $d = 1$ this is the familiar tent map of the unit interval. The case $d = 2$ is illustrated in Figure 1.

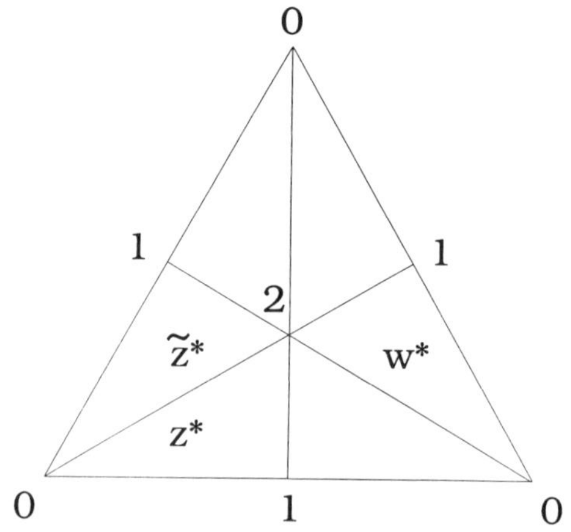

Figure 1

A d dimensional simplex of K^* is obtained by a permutation π on $\{0,\ldots,d\}$ with vertices $b([\pi 0]), b([\pi 0, \pi 1]), \ldots, b([\pi 0, \ldots, \pi d]) = b(\sigma)$. In particular, there are $(d+1)!$ simplices in K^* of dimension d. Since g maps these vertices to $0, 1, \ldots, d$, respectively, it follows that g is nondegenerate and maps each d simplex of K^* to the unique d simplex σ of K. Thus, the d skeleton basic set B^* consists of all of the d simplices of K^*, i.e. $B^* = S^d(K^*)$. So B^* is a polyhedral basic set with $|B^*| = X$. By Corollary 4.25 $h^+(B^{*+}_{G^*}) = X$ and by Lemma 4.2 $h(B^*_{G^*}) = X_g$. So Theorem 4.15b implies that $h^+ : B^{*+}_{G^*} \to X$ and $h : B^*_{G^*} \to X_g$ are almost homeomorphisms. By Theorem 5.6 the entropy of g on X is that of the full shift, $\ln((d+1)!)$. All of the simplices of B^* have the same Lebesgue measure and so the Lebesgue measure λ_B is the Parry measure μ_{B^*}.

Because the associated G basic set B consists of a single simplex we have an interesting special result.

1 Proposition. *With $g : K^* \to K$ a simplicial dynamical system, let B^* be a basic set for G^* whose associated basic set B for G contains exactly one element. Then*

$$G^* = B^* \times B^* \tag{7.1}$$

and $s^+_{G^*}$, s_{G^*} are the full shifts on $B^{*+}_{G^*} = (B^*)^{\mathbf{Z}_+}$ and $B^*_{G^*} = (B^*)^{\mathbf{Z}}$, respectively. The continuous map g on $h^+(B^{*+}_{G^*})$ and the homeomorphism s_g on $h(B^*_{G^*})$ satisfy the Shadowing Property (see Appendix Chapter 11).

Proof. If $z^* \in K^*$ then by (4.10)

$$z^* \in B^* \Leftrightarrow j(z^*) = \sigma = g(z^*) \tag{7.2}$$

where $B = \{\sigma\}$. In particular, if $z_1^*, z_2^* \in B^*$ then $z_2^* \in G^*(z_1^*)$, proving (7.1). It is then clear that $s^+_{G^*}$ and s_{G^*} are full shifts on $B^{*+}_{G^*}$ and $B^*_{G^*}$.

Now let $\eta \in X^{\mathbf{Z}}$ or $X^{\mathbf{Z}_+}$ such that $d_K(\eta_{i+1}, g(\eta_i)) \le \epsilon$ and $\eta_i \in h(B^*_{G^*})$ for all $i \in \mathbf{Z}$ (or $\eta_i \in h^+(B^{*+}_{G^*})$ for all $i \in \mathbf{Z}_+$). We can then choose $z_i^* \in B^*$ such that $\eta_i \in z_i^*$ for all i. By (7.1) this defines $\mathbf{z}^* \in B^*_{G^*}$ (resp. $\mathbf{z}^* \in B^{*+}_{G^*}$) and so by Theorem 3.3 η is $(d/\theta)\epsilon$ shadowed by $h(\mathbf{z}^*)$ (resp. by $h^+(\mathbf{z}^*)$). □

While it follows that g on X satisfies the Shadowing Property it is not true that g is expansive. We use the three simplices labeled z^*, $\tilde{z}^*$, w^* in Figure 1.

Observe first that with $v = b([0,1])$ the sequence $\{(\overline{g}_{z^*})^i(v) : i = 0,1,\ldots\}$ is in the one simplex $[0,v]$ of K^* and converges to the vertex 0. So given $\epsilon > 0$ we can choose $n > 4$ such that $d_K(0, (\overline{g}_{z^*})^{n-4}(v)) < \epsilon$. So we can define the ϵ chain η for g by:

$$\eta_0 = 0,\ \eta_1 = (\overline{g}_{z^*})^{n-4}(v),\ \eta_2 = (\overline{g}_{z^*})^{n-5}(v), \ldots$$

$$\eta_{n-3} = v, \quad \eta_{n-2} = 1, \quad \eta_{n-1} = 0$$

$$\eta_{k+n} = \eta_k \text{ for } k \in \mathbf{Z}. \tag{7.3}$$

Now let ω be an arbitrary element of $\{z^*, \tilde{z}^*\}^{\mathbf{Z}}$. Define $\mathbf{z}^*(\omega) \in B^*$

$$z^*(\omega)_{in} = \tilde{z}^*$$

$$z^*(\omega)_{1+in} = \ldots = z^*(\omega)_{n-3+in} = z^*$$

$$z^*(\omega)_{n-2+in} = w^*$$

$$z^*(\omega)_{n-1+in} = \omega_i \text{ for all } i \in \mathbf{Z}. \tag{7.4}$$

Since $0 \in z^* \cap \tilde{z}^*$ we have

$$\eta_i \in z^*(\omega)_i \text{ for all } i \in \mathbf{Z}. \tag{7.5}$$

By Theorem 3.3 again it follows that $\xi(\omega) \equiv h(\mathbf{z}^*(\omega))$ 6ϵ shadows η ($d = 2$ and $\theta = 1/3$ in this case). It is easy to check that for no $i \in \mathbf{Z}$ can $\xi(\omega)_i$ lie in a one simplex of K^* because the g orbit of such points eventually remain in $[0,1]$ and so cannot enter w^* excepting the case that the orbit is eventually at the fixed point 0. It follows that $\omega_i \neq \tilde{\omega}_i$ implies $\xi(\omega)_{n-1+in} \neq \xi(\tilde{\omega})_{n-1+in}$. So the map $\omega \mapsto \xi(\omega)$ is injective and the uncountable image set of points all shadow one another in X_g with distance at most 12ϵ.

Since g and s_g are not expansive it follows from Proposition 11.15 of the Appendix that the equivalence relations $(h^+)^{-1} \circ (h^+)$ and $h^{-1} \circ h$, while subshifts of $(B^* \times B^*)^+_{G^* \times G^*}$ and $(B^* \times B^*)_{G^* \times G^*}$ respectively, are not of finite type. By Proposition 3.4 these relations are projections of subshifts defined using $(DK^*) \cap (B^* \times K^* \times B^*)$ (see (4.26)) and so they are so-called sofic systems.

On the boundary $\partial\sigma$ g has image the $d-1$ dimensional base simplex on which g restricts to the $d-1$ dimensional tent map. □

Example (2) (Space Filling Curve).

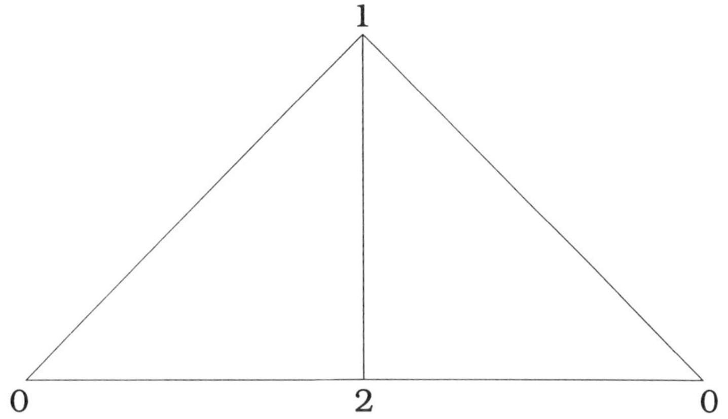

Figure 2

The subdivision K^* is not proper and so $g : K^* \to K$ is not a simplicial dynamical system. However, if we use g to pull back the subdivision K^* of K we obtain K^{**} a subdivision of K^* such that $g : K^{**} \to K^*$ is simplicial. Repeating the process we obtain K^{***} such that $g : K^{***} \to K^{**}$ is simplicial. In this case, K^{***} is a proper subdivision of K and so $g^3 : K^{***} \to K$ is a simplicial dynamical system. Thus, $g : K^* \to K$ is a generalized simplicial dynamical system in the sense of Chapter 6. Furthermore, if we use not the l^1 metric d_K on σ but the Euclidean metric illustrated in Figure 2 with a right angle at vertex 2, then each local inverse $\overline{g}_{z^*}$ is a contracting similarity map.

Let $g_0 : K_0^* \to K_0$ be the one dimension tent map on $X = [0, 1]$. We define the finite set map $f : K_0^* \to K^*$ by

$$f(0) = 0, \ f(1) = 1, \ f(b([0,1])) = [b([0,1]), 2]$$

$$f([0, b([0,1])]) = [0, b([0,1]), 2]$$

$$f([b([0,1]), 1]) = [b([0,1]), 2, 1]. \tag{7.6}$$

Since f satisfies (3.58) and (3.59) mapping g_0 to g, Theorem 3.7 applies and there exists a continuous map $\varphi : [0,1] \to [0,1,2]$ such that

$$g \circ \varphi = \varphi \circ g_0. \tag{7.7}$$

Letting B_0^* be the 1 skeleton basic set for G_0^* in K_0^* we clearly have the commutative diagram:

$$\begin{array}{ccc} B_0^{*\mathbf{Z}_+} & \xrightarrow{\;f\;} & B^{*\mathbf{Z}_+} \\ \Big\downarrow h_0^+ & & \Big\downarrow h^+ \\ [0,1] & \xrightarrow[\;\varphi\;]{} & [0,1,2] \end{array} \tag{7.8}$$

Because f is a homeomorphism and h^+, h_0^+ are almost homeomorphisms we see that φ is surjective. The Lebesgue measures λ_0 on $[0,1]$ and λ on $[0,1,2]$ are the images under h_0^+ and h^+ of the Parry measure for the 2 shift, i.e. the Bernoulli shift with probabilities $\{\frac{1}{2}, \frac{1}{2}\}$. Hence,

$$\varphi_* \lambda_0 = \lambda. \tag{7.9}$$

Recall that there is a one parameter family of isomorphs of the tent map obtained by cutting the interval into lengths $\{p, 1-p\}$ instead of at the midpoint. Over these we get a one parameter family of right triangles whose upper vertices lie on the half-circle with diameter $[0,1]$. We thus obtain the one parameter family of space filling curves discussed by Mandelbrot and Jaffard (1997). In these cases, the measure λ_0 on $[0,1]$ and λ on $[0,1,2]$ corresponds to the Bernoulli 2 shift with probabilities $\{p, 1-p\}$.

There is a higher dimensional analogue of all this. Let $g_0 : K_0^* \to K_0$ be the d dimensional tent map of Example (1). Let K^* triangulate the $d+1$ simplex as the cone on K_0^*. That is, there is one new vertex $\tilde{d}$ and the remaining simplices of K^* consist of simplices of K_0^* and simplices of K_0^* with $\tilde{d}$ adjoined to the vertex set. Define $g : K^* \to K$ by $g(b(z)) = k$ if z is a simplex of K_0 with dimension $z = k < d$. Let $g(b(\sigma)) = \tilde{d}$ instead of d and let $g(\tilde{d}) = d$. As in the above case with $d = 1$, g is a generalized simplicial dynamical system and there is a continuous map from σ onto $[\sigma, \tilde{d}] = |K|$ which satisfies (7.7) and (7.9). □

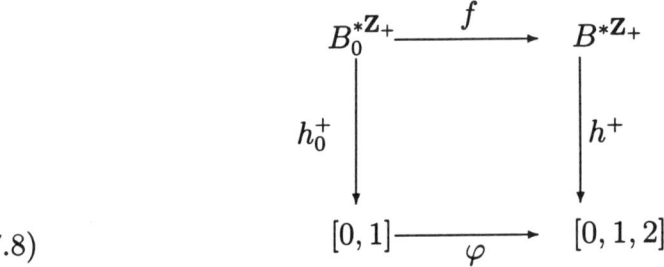

Example 3. Let π be a permutation of $\{0, 1, 2\}$.

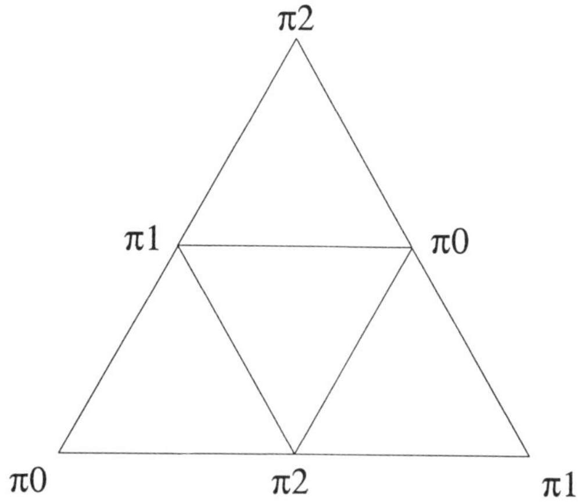

Figure 3

The map g is nondegenerate and so B^* associated with $B = \{\sigma\}$ is the polyhedral basic set consisting of the four 2 simplices of K^*. As in Examples (1) and (2) $s^+_{G^*}$ and s_{G^*} are full shifts and $h^+ : B^{*\mathbf{Z}_+} \to X$ and $h : B^{*\mathbf{Z}} \to X_g$ are almost homeomorphisms. The entropy of g is $\ln 4$.

On the boundary, triangulated by the 1 skeleton $\overline{S}^1(K)$, G relates the three 1 simplices to distinct pairs of 1 simplices. It easily follows that the three 1 simplices form a single 1 skeleton basic set B^1 for G which is thus associated to the G^* basic set B^{*1} consisting of the six 1 simplices of K^* in the boundary $\#j(z) = \#g(z) = 2$ for each 1 simplex z in B^1. The boundary is $h^+(B^{*1+}_{G^*})$ for this polyhedral 1 skeleton basic set B^{*1}. The restriction of g to the boundary thus has entropy $\ln 2$.

The details of the map do depend on π. For example, if $\pi 0 = 0$ and π interchanges 1 and 2 then g on the boundary is conjugate to the map $t \to 2t$ on the reals mod 1, i.e. the group $\mathrm{R}/\mathbf{Z}$. □

Example (4) (The Onishi Example). As was illustrated by the tent map examples discussed at the end of Chapter 5, isomorphs of g may have different entropy with respect to the same invariant Lebesgue measure.

In that case the maximum entropy among invariant measures, i.e. the topological entropy, does occur among these isomorphs. That this need not always happen is shown by the following example constructed by H. Onishi. Again let π be a permutation of $\{0, 1, 2\}$.

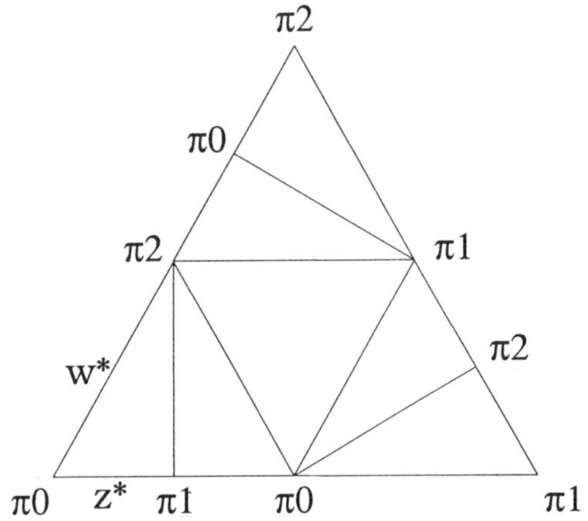

Figure 4

Because g is nondegenerate the 2 skeleton basic set B^* consists of the seven 2 simplices of K^*. B^* is polyhedral, $s^+_{G^*}$ and s_{G^*} are full shifts and $h : B^{*\mathbf{Z}} \to X_g$, $h^+ : B^{*\mathbf{Z}_+} \to X$ are almost homeomorphisms. The map g on X has entropy ln 7. The Parry measure μ_{B^*} on X is the projection via h^+ of the Bernoulli measure with equal weight $1/7$ on each letter of the alphabet B^*.

Observe that K^* is a subdivision of the subdivision illustrated by Figure 3 which we will now call $\tilde{K}$. $\tilde{K}$ has a single "central" 2 simplex and three "outside" 2 simplices each of which we label $\sigma_0, \sigma_1, \sigma_2$ by the vertex of K which it contains. Suppose we replace $\tilde{K}$ be an isomorph obtained by subdividing $[0, 1]$ into pieces of length $(x, 1-x)$, $[1, 2]$ into pieces of length $(y, 1-y)$ and $[2, 0]$ into pieces of length $(z, 1-z)$. Since each of the outside triangles contains a $60°$ angle at its vertex and because Lebesgue measure of the entire equilateral triangle is normalized to area 1, we obtain

$$\lambda(\sigma_0) = x(1-z)$$

$$\lambda(\sigma_1) = y(1-x)$$

(7.10) $$\lambda(\sigma_2) = z(1-y).$$

Now suppose that the three areas are equal and that $x \geq y$. Then $1 - x \leq 1 - y$ and so $\lambda(\sigma_1) = \lambda(\sigma_2)$ implies $y \geq z$. Similarly, $\lambda(\sigma_2) = \lambda(\sigma_0)$ then implies $z \geq x$. As $x \geq y \geq z \geq x$ we have $x = y = z$. The common value for the area is $x(1-x)$ which is maximum when $x = 1/2$, the original symmetric picture of Figure 3. We have thus shown that if the three outside triangles have equal area then the central triangle, which we will denote by σ_4 has area at least 1/4, i.e.

$$\lambda(\sigma_0) = \lambda(\sigma_1) = \lambda(\sigma_2) \Rightarrow$$

(7.11) $$\lambda(\sigma_4) \geq \frac{1}{4}.$$

Now return to the subdivision K^*. The central simplex $\sigma_4 \in K^*$ but for $k = 0, 1, 2$ the simplex σ_k is cut into two 2 simplices σ_k' and σ_k'' where σ_k' contains the vertex k. The seven simplices $\{\sigma_4, \sigma_0', \sigma_0'', \sigma_1', \sigma_1'', \sigma_2', \sigma_2''\}$ provide a partition of σ in the sense of Lebesgue measure λ because the intersection of any two has measure zero. For any subdivision isomorph of K^*, the Lebesgue measure λ on σ is the image under the appropriate h^+ of the Bernoulli measure on $B^{*\mathbf{Z}_+}$ with probabilities $\{\lambda(\sigma_4), \lambda(\sigma_0'), \ldots\}$. The entropy of λ with the isomorph of g is just the entropy of the partition. For example, in Figure 4 itself $\lambda(\sigma_4) = 1/4$ and each of the remaining simplices has area 1/8. Thus, the entropy of g with respect to λ is

(7.12) $$H = \frac{3}{4}\ln 8 + \frac{1}{4}\ln 4$$

which lies strictly between ln 6 and ln 7. By (7.11) no isomorph of K^* yields a partition with all seven areas equal. In fact, it is not hard to show that among partitions such that $\lambda(\sigma_0) = \lambda(\sigma_1) = \lambda(\sigma_2)$ H of (7.12) is the maximum entropy value. It follows that the entropy for each isomorph of K^* is less than ln 7. In fact, the supremum of such entropies is less than ln 7. This follows because, by compactness, the supremum is achieved among the isomorphs of K^*. (I conjecture that H is the supremum.) For if a sequence of isomorphs has no subsequence convergent with respect to the metric of Proposition 3.9d then at least one of the areas must degenerate to zero and so we are left with a partition of the number 1 into fewer than seven positive values. The associated entropy is at most ln 6 which is less than H.

Now look at the case with $\pi 0 = 0$, $\pi 1 = 2$ and $\pi 2 = 1$. Let $X_0 = [0,1] \cup [0,2]$ and $g_0 : K_0^* \to K_0$ be the restriction of g to this subset of σ. The six 1 simplices of K_0^* form a polyhedral 1 skeleton basic set B_0^* whose associated G_0 basic set $B_0 = \{[0,1],[0,2]\}$. The relation G_0 on B_0 is just the map of period 2 exchanging $[0,1]$ and $[0,2]$. This is the first level of complexity beyond the singleton case described in Proposition 1. Nonetheless we will show that g_0 on $X_0 = h_0^+(B_{0G_0^*}^{*+})$ does not satisfy the Shadowing Property.

We use the 1 simplices labeled z^* and w^* in Figure 4. Let $y_0 = b([0,1])$, $y_{-1} = \overline{g}_{w^*}(y_0)$, $y_{-2} = \overline{g}_{z^*}(y_{-1})$, etc. Thus we obtain a g chain indexed by the negative integers with $g(y_0) = 0$ and $\mathrm{Lim}_{n\to\infty} y_{-n} = 0$. Thus, given $\epsilon > 0$ there is a positive integer n such that $d_K(y_{-2n+2}, 0) < \epsilon$. Define

$$\eta_0 = y_{-2n+2}, \ldots, \eta_{2n-2} = y_0,\ \eta_{2n-1} = 0,\ \eta_{2n} = 0$$

$$\eta_{k+(2n+1)} = \eta_k \text{ for all } k \in \mathbf{Z}. \tag{7.13}$$

Because the jump from $0 = g(0)$ to η_0 has distance at most ϵ this defines an ϵ chain for g periodic with period $2n+1$. In particular, $\eta_i \in [0,1]$ for even i in the interval $0 \le i \le 2n$ implies $\eta_i \in [0,1]$ for odd i in the interval $2n+1 \le i \le 2n+(2n+1)$. Now let $\xi \in X_g$. I claim it cannot happen that

$$d_K(\xi_i, \eta_i) < \frac{1}{2} \text{ for all } i \in \mathbf{Z}. \tag{7.14}$$

In fact, we need only consider those i's congruent to $2n-2$ mod $2n+1$. For such values $\eta_i = b([0,1])$ and so $\xi_i \neq 0$ by (7.13). Since 0 is a fixed point for g we have $\xi_i \neq 0$ for any $i \in \mathbf{Z}$. Since $g((0,1]) = [0,2]$ and $g((0,2]) = [0,1]$ it follows that either $\xi_i \in (0,2]$ for all even i or $\xi_i \in (0,2]$ for all odd i. But (7.14) implies $\xi_i \in (0,1]$ for all i congruent to $2n-2$ mod $2n+1$. In particular, ξ_{2n-2} and $\xi_{4n-1} \in [0,1]$. This contradiction implies (7.14) does not hold and so no true orbit δ shadows η for any $\delta < 1/2$. □

In the remaining examples there are simplices on which g is degenerate.

We shade these in the accompanying figures. In each one B^* is tattered and so by Proposition 5.12 $h^+(B^*_{G^*})$ is a closed nowhere dense subset of σ with 2 dimensional Lebesgue measure zero. In each case we denote by F the closed invariant subset $h^+(B^*_{G^*})$.

Example (5). This map g fixes the vertices 0,1,2 of K. The 2 skeleton basic set $B = \{\sigma\}$ for G is associated with $B^* = \{z^*\}$ where z^* is the unique nondegenerate 2 simplex in the center. h^+ maps the singleton set $B^{*+}_{G^*}$ to $\bigcap_{m=0}^{\infty} g^{-m}(z^*) = \{b\}$ where b is the barycenter of σ. So B^* is an interior basic set,

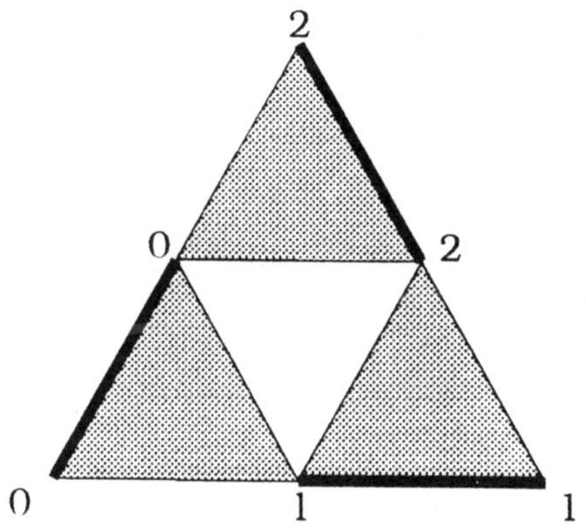

Figure 5

The fixed point b is a repellor for g whose dual attractor is the boundary of σ (see Akin (1993) Chapter 3).

On the boundary each face is invariant and is a separate 1 skeleton basic set, e.g. $B_1 = \{[0,1]\}$ is associated with $B_1^* = \{[0, b([0,1])]\}$. These are not interior basic sets as $\bigcap_{m=0}^{\infty} g^{-m}([0, b([0,1])]) = \{0\}$ and the carrier of 0 is 0 itself.

Thus, the center of g, the closure of the recurrent points, is the fixed point set $\{0,1,2,b\}$. If x is not in the center then $g^i(x) \in \{0,1,2\}$ for some positive i. For example, if x is in the interior of $[0,1]$ then $g^i(x)$ is eventually equal to 1. In particular, $1 \in \mathcal{C}g(0)$ and the entire boundary is a single basic set for g. □

Example (6). Here again g fixes the vertices. $B = \{\sigma\}$ is associated with $B^* = \{z, \tilde{z}\}$, the two nondegenerate 2 simplices. $|B^*| = z^* \cup \tilde{z}^*$ is a neighborhood of 2 with $\{2\} = \cap_{m=0}^{\infty} g^{-m}(|B^*|)$. So $\{2\}$ is a repellor for g and B^* is not an interior basic set.

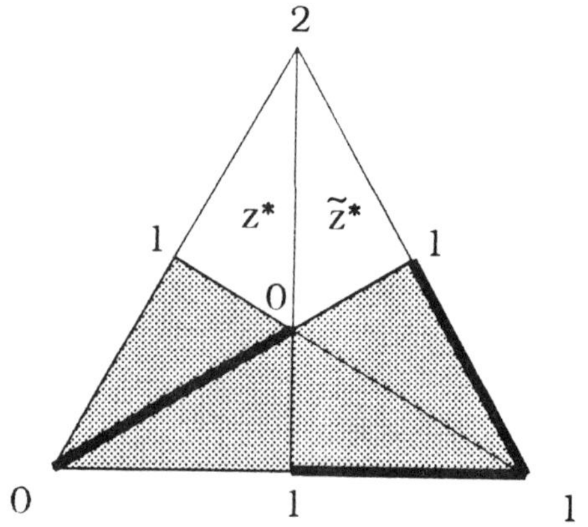

Figure 6

The attractor dual to $\{2\}$ is the 1 simplex $[0,1]$ on which $\tilde{B}^* = \{[0, b([0,1])]\}$ is a noninterior 1 skeleton basic set for G^* with h^+ mapping $\tilde{B}^{*+}_{G^*}$ to $\{0\}$. The other 1 skeleton basic set for G^* is $\{[2, b([1,2])]\}$ which is also not interior. The fixed point 1 has neighborhoods U such that $g^2(U) = \{1\}$. So $\{1\}$ is an attractor whose dual repellor is easily seen to be the union of the increasing sequence of compact connected sets $\{g^{-m}(0) : m = 0, 1, \ldots\}$ together with its limit point $\{2\}$. Thus, $\{2\}$, $\{0\}$, $\{1\}$ are three distinct basic sets linearly ordered by $\mathcal{C}g$.

Finally, notice that while g has entropy zero (the center is a finite set), the shift s_{G^*} on $B^*_{G^*}$ is the full shift on two symbols and so has entropy log 2. Thus, the maximum in (5.69) cannot be extended over the noninterior basic sets. □

In the examples which follow some of the basic set images are not convex subsets of σ. To study them we use the Barnsley contraction operator $\overline{g}_{B^*}$ on $\Pi_{z \in B} C'(z)$ defined by (4.57). Notice that when B consists of a single simplex z then $\overline{g}_{B^*}$ is an example of an iterated function system rather than a generalization thereof.

Example (7). The subdivision is isomorphic to that of Figure 4. On the boundary g is conjugate to the map $t \mapsto 3t$ on the reals mod 1. The three 1 simplices of $\partial\sigma$ form a G basic set $\tilde{B}$ whose associated G^* basic set $\tilde{B}^*$ consists of the nine 1 simplices of K^* in $\partial\sigma$. $\tilde{B}^*$ is an polyhedral basic set with $h^+(\tilde{B}^{*+}_{G^*}) = \partial\sigma$. Because g is nondegenerate on the boundary $\{x \in \partial\sigma : g^i(x) = 0 \text{ for some } i \in \mathbf{Z}_+\}$ is countable and is the complement of $h^+(\tilde{B}^{*+}_{G^*} \cap \mathrm{Int}^+_1)$.

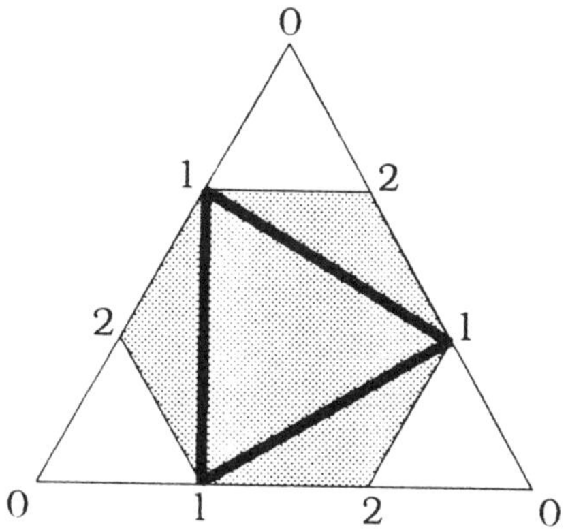

Figure 7

Associated with $B = \{\sigma\}$ is $B^* = \{z^{*0}, z^{*1}, z^{*2}\}$ the three nondegenerate simplices in the corners. $\overline{g}_{B^*}$ is a continuous map on $C'(\sigma)$. We will write α for $\alpha_\sigma \subset \sigma$. Beginning with $\alpha = \{0\}$, $\overline{g}_{B^*}(\alpha) = \{0, 1, 2\}$ which contains α. As $\overline{g}_{B^*}(\alpha) \supset \alpha$ it follows that $\alpha^* = h^+(B^{*+}_{G^*})$ is the closure of the union of the increasing sequence $\{\overline{g}^m_{B^*}(\alpha)\}$. Since $\overline{g}^3_{B^*}(\alpha)$ meets $(\sigma)^\circ$ it follows that the basic set image $h^+(B^{*+}_{G^*})$, which we denote F, meets $(\sigma)^\circ$ and so B^* is an interior basic set. Proposition 4.23(a) applies to show that h^+ is a homeomorphism from $B^{*+}_{G^*} = (B^{*+})^{\mathbf{Z}_+}$ onto F. Thus, F is a Cantor space (i.e. a perfect, compact metric space). $F \cap [0, 1]$ is the classical Cantor set. So the basic sets images F and $\partial\sigma$ meet but neither includes the other. □

Example (8). $\{x \in \partial\sigma : g^i(x) = 0 \text{ for some } i \in \mathbf{Z}_+\}$ consists of 0,2 and a sequence converging to 1 in the 1 simplex $[1, b[1,2]]$ of K^*. For every other point x in $\partial\sigma$ $g^i(x)$ eventually equals 1. The two 1 skeleton basic sets for G^*, $\{[1, b[1,2]]\}$ and $\{[0, b[0,1]]\}$ are noninterior.

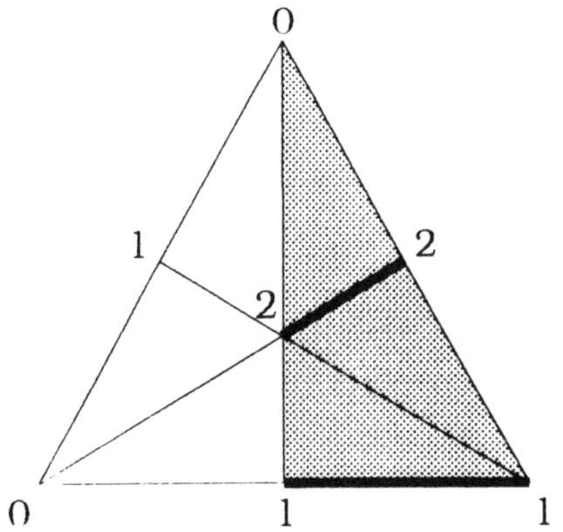

Figure 8

The 2 skeleton basic set B^* associated with $B = \{\sigma\}$ consists of the three nondegenerate 2 simplices of K^*. $\overline{g}_{B^*}(\{0\}) = \{0, 2\} \supset \{0\}$ and $(\overline{g}_{B^*})^2(\{0\}) = \{0, 2, b\}$ where $b = b(\sigma)$. It follows that the limit set $\alpha^* \supset \{0, 2, b\}$. For each of the three simplices z^* in B^*, $\overline{g}_{z^*}(\alpha^*) \supset \{\overline{g}_{z^*}(2)\} = \{b\}$. By Proposition 4.23b applied with $\alpha = \{b\}$ it follows that $\alpha^* = h^+(B^{*+}_{G^*}) = F$ is a connected compact set containing the interior point b. F is a closed subset of the left half of σ, the triangle $[0, 2, b[0,1]] = |B^*|$. Furthermore,

$$F \cap \partial\sigma = \{0, 2\}. \tag{7.15}$$

That $F \cap \partial\sigma \supset \{0, 2\} = \overline{g}_{B^{*2}}(\{0\})$ is clear from monotonicity and (5.20). On the other hand, for every other point x in $[0,2] \cup [0, b[0,1]]$ $g^i(x) = 1$ for $i \in \mathbf{Z}_+$ large enough and so x does not lie in the invariant set F because 1 does not.

We have seen that there is a sequence of points converging to 1 which are eventually mapped to 0 and so into F by iterating g. On the other hand, by Proposition 4.18a for every neighborhood of $x \in F = h^+(B^{*+}_{G^*})$ contains points $\tilde{x}$ whose endset is the vertex $\{1\}$. It follows that F and $\{1\}$, the two maximal basic set images whose union is the center, both lie in the same basic set for g. Now for any point $x \in X$ there exists $\xi \in X_g$ such that $\xi_0 = x$, i.e. $\pi_0 : X_g \to X$ is surjective because g is surjective. $\alpha s_g(\xi) \cup \omega s_g(\xi) \subset h_{G^*}(B^*_{G^*}) \cup \{1\}$ where $1 \in X_g$ denotes the sequence constant at the point 1. It follows that for every $x \in X$ $\mathcal{C}g(x) \supset \omega g(x) = \pi_0(\omega s_g(\xi))$ and $\mathcal{C}g^{-1}(x) \supset \pi_0(\alpha s_g(\xi))$ both meet $F \cup \{1\}$. Thus, all of $X = \sigma$ consists of a single basic set, i.e. g is chain transitive on X (see Akin (1993) Theorem 4.12).

We now prove that the point $b[0,1]$ is nonwandering. Since this point is not in the center, $F \cup \{1\}$, it follows that for this example the closure of the recurrent points, the center, is a proper subset of the nonwandering set, compare Proposition 4.13b.

With $x = b[0,1]$ let $z^*_0 = [0, b[0,1], b]$, so that $x \in z^*_0$, and $z^*_i = [1, b[1,2]]$ so that $1 = g^i(x) \in z^*_i$ for all $i > 0$. This defines $\mathbf{z}^* \in K^{*+}_{G^*}$. Observe that for $B^* = \{[1, b[1,2]]\}$, a basic set for G^*, $G^{*2}(B^*)$ contains the vertex $b[0,1]$. It follows from Proposition 4.18a that for every neighborhood U of x and positive integer N there exist $\tilde{x} \in U$ and $\tilde{N} \geq N$ such that $g^{\tilde{N}}(\tilde{x}) = x$. Hence, x is nonwandering. It is possible to prove similarly that all of the points of $[0,1] \cup [0,2]$ are nonwandering. □

Example 9. On $[0,1]$ g restricts to the dimension 1 tent map, with $\{[0, b([0,1])], [b([0,1]), 1]\}$ an interior 1 skeleton G^* basic set.

$$\tilde{B}^* = \{[b([0,2]), 2], [1, b([1,2])]\}$$

is an interior 1 skeleton basic set for G^* with $\{[0,2],[1,2]\}$ the associated G basic set. $\tilde{B}^{*+}_{G^*}$ consists of two points whose image under h^+ is the periodic orbit $\{x_0, x_1\}$ for g with

$$x_0 = \frac{1}{3}0 + \frac{2}{3}2$$

$$x_1 = \frac{2}{3}1 + \frac{1}{3}2. \tag{7.16}$$

The vertex 2 is a fixed point.

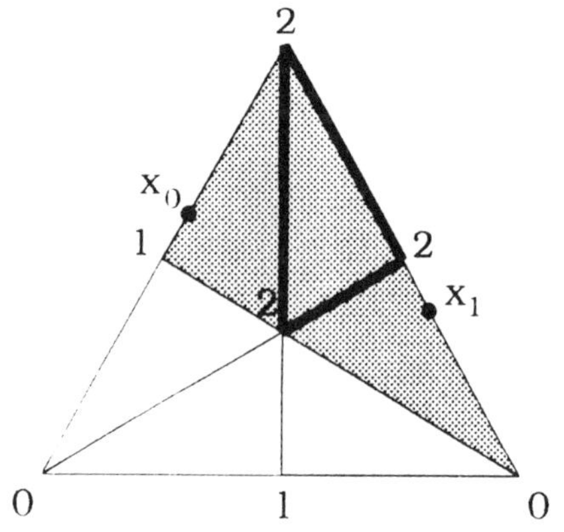

Figure 9

It is easy to construct a neighborhood U of 2 such that $g^2(U) = \{2\}$. So $\{2\}$ is an attractor.

The associated 2 skeleton basic set B^* for G^* consists of the three non-degenerate simplices. Let $\alpha = [0,1] \subset \sigma$. $\overline{g}_{B^*}(\alpha) = [0,1] \cup [0, b([0,2])] \supset \alpha$ and $\overline{g}^2_{B^*}(\alpha)$ meets the interior $(\sigma)^\circ$. Consequently B^* is an interior basic set and $F = h^+(B^{*+}_{G^*}) = \alpha^*$, is a connected compact set by Proposition 4.23c.

By Proposition 4.18a every point x of F can be arbitrarily closely approximated by points $\tilde{x}$ such that $g^i(\tilde{x}) = 2$ for i large enough. For let $\mathbf{z}^* \in B^{*+}_{G^*}$ with $h^+(\mathbf{z}^*) = x$ and N be an arbitrarily large positive integer. Define $\tilde{z}^*_i = z^*_i$ for $i \leq N$ and $\tilde{z}^*_i = 2$ for $i > N$. This is a G^* chain with $\tilde{x} = h^+(\tilde{\mathbf{z}}^*)$ close to x when N is large. Furthermore $g^{N+1}(\tilde{x}) = h^+(s^{N+1}_{G^*}(\tilde{\mathbf{z}}^*)) = 2$. If, instead, we let $\tilde{z}^*_{N+1} = [b([0,2]), 2]$, $\tilde{z}^*_{N+2} = [1, b([1,2])]$ and then continue alternating between these values we have that $g^{N+1}(\tilde{x}) = x_0$ and so close to x are points which eventually hit the two cycle $\{x_0, x_1\}$.

On the other hand, on the boundary there are points $\tilde{x}$ arbitrarily close to x_0 and x_1 such that $g^i(\tilde{x}) \in [0,1]$ for i large enough.

It follows that while the center for g consists of $F \cup \{x_0, x_1\} \cup \{2\}$,

$F \cup \{x_0, x_1\}$ is contained in a single basic set. It is the repellor dual to the attractor $\{2\}$. □

8. PL Roundoffs of a Continuous Map.

For a complex K on X, a subset K_1 is a subcomplex when all the faces of a simplex of K_1 are also in K_1. We will say of the support, $|K_1| = \cup\{z \in K_1\}$, of a subcomplex that it is a subset of X *triangulated by* K. So $Y \subset X$ is triangulated by K when it is a union of closed simplices of K, or, equivalently, when

$$x \in Y \Rightarrow q(x) \subset Y \tag{8.1}$$

where $q(x)$ is the carrier of x in K as defined in (2.17). If Y is triangulated by K and K^* is any subdivision of K then Y is triangulated by K^* as well.

Using the open star, St, of a simplex defined in (2.13), we define for any subset Y of X the *star neighborhood*, $N(Y, K)$. and the *open star neighborhood*, $N^\circ(Y, K)$, of Y for K:

$$N(Y, K) = \cup\{z \in K : z \cap Y \neq \emptyset\}.$$

$$\begin{aligned} N^\circ(Y, K) &= \cup\,\{(z)^\circ \;: z \in K \text{ and } z \cap Y \neq \emptyset\} \\ &= \cup_{x \in Y} St(q(x)). \\ C(Y, K) &= X \backslash N^\circ(Y, K) \\ &= \cup\{z \in K : z \cap Y = \emptyset\}. \end{aligned} \tag{8.2}$$

$N^\circ(Y, K)$ is an open neighborhood of Y with closure $N(Y, K)$. $N(Y, K)$ and $C(Y, K)$ are unions of closed simplices of K and so are triangulated by K.

If $Y = \emptyset$ then, as they are empty unions, $N(Y, K) = N^\circ(Y, K) = \emptyset$.

If Y is triangulated by K then by (8.1) and (8.2)

$$N^\circ(Y, K) = \cup_{z \subset Y} St(z). \tag{8.3}$$

Of special interest is the star neighborhood for a first derived subdivision of K, e.g. the barycentric subdivision of K.

1 Lemma. *Assume that K' is a first derived subdivision of K. If Y_1 and Y_2 are subsets triangulated by K then*

$$N^\circ(Y_1, K') \cap N^\circ(Y_2, K') = N^\circ(Y_1 \cap Y_2, K'). \tag{8.4}$$

In particular, if Y_1 and Y_2 are disjoint then the star neighborhoods for K' are disjoint.

For example, if $z_1, z_2 \in K$ then

$$N^\circ(z_1, K') \cap N^\circ(z_2, K') = N^\circ(z_1 \cap z_2, K'). \tag{8.5}$$

If z_1 and z_2 have no common face in K then $N^\circ(z_1, K')$ and $N^\circ(z_2, K')$ are disjoint.

Proof. Let the carrier of x in K' be the simplex $[b(z_0), \dots, b(z_k)]$ with z_{i-1} a proper face of z_i in K for $i = 1, \dots, k$. If $x \in N^\circ(Y_1, K')$ then, since Y_1 is triangulated by K', (8.3) implies that $b(z_{i_1}) \in Y_1$ for some i_1 between 0 and k. Since Y_1 is triangulated by K, (8.1) implies that z_{i_1}, the carrier of $b(z_{i_1})$, is a subset of Y_1. So if x is in $N^\circ(Y_2, K')$ as well then $z_{i_2} \subset Y_2$ for some i_2 between 0 and k. Without loss of generality we can assume $i_1 \le i_2$ so that $z_{i_1} \subset z_{i_2} \subset Y_2$. Thus, $z_{i_1} \subset Y_1 \cap Y_2$ and so $x \in N^\circ(Y_1 \cap Y_2, K')$. □

If L and K are simplicial complexes then a function $s : L \to K$ is called an *incidence preserving map* if

$$z_1 < z_2 \text{ in } L \Rightarrow s(z_1) < s(z_2) \text{ in } K. \tag{8.6}$$

All simplicial maps are incidence preserving, but as we have seen in Example (2) of Chapter 7 the converse need not hold.

If $f : |L| \to |K|$ is a continuous map and $s : L \to K$ is an incidence preserving map then we say that s *approximates* f if for every $z \in L$

$$f(z) \subset N^0(s(z), K') \tag{8.7}$$

where K' is the barycentric subdivision of K.

Observe that for any incidence preserving map $s : L \to K$ the set of continuous maps $f : |L| \to |K|$ approximated by s is open in the uniform topology. That is, if s approximates f then there exists $\epsilon > 0$ so that $f_1 : |L| \to |K|$ continuous with $d_K(f(x), f_1(x)) \le \epsilon$ for all $x \in |L|$ implies s approximates f_1.

2 Proposition. *Let $f : Y \to X$ be a continuous map between polyhedra. If L and K are simplicial complexes with $|L| = Y$ and $|K| = X$ then we say that L is* fine *for f and K if the following equivalent conditions hold:*

(1) There exists an incidence preserving map $s : L \to K$ *which approximates* f.

(2) For every $z \in L$ *there exists* $w \in K$ *such that* $f(z) \subset N^\circ(w, K')$.

If L *is fine for* f *then for each* $z \in L$ *define* $s_f(z)$ *to be the simplex* w *of smallest dimension such that* $f(z) \subset N^\circ(w, K')$, $s_f : L \to K$ *is an incidence preserving function approximating* f *and furthermore for all* $z \in L$ *and* $w \in K$:

$$f(z) \subset N^\circ(w, K') \Rightarrow s_f(z) < w. \tag{8.8}$$

In particular, if $s : L \to K$ *is an incidence preserving map approximating* f *then:*

$$s_f(z) < s(z) \text{ for all } z \in L. \tag{8.9}$$

Moreover, if L_0 *is a subcomplex of* L *and* $f_0 : L_0 \to K$ *is a simplicial map such that the restriction of* f *to* $|L_0|$ *is the p.l. map* $f_0 : |L_0| \to |K|$ *then*

$$s_f(z) = f_0(z) \text{ for all } z \in L_0. \tag{8.10}$$

Proof. That (1) implies (2) is obvious. By Lemma 1, $f(z) \subset N^\circ(w_1, K')$ and $f(z) \subset N^\circ(w_2, K')$ implies $f(z) \subset N^\circ(w_1 \cap w_2, K')$. Hence, $\{w \in K : N^\circ(w, K') \supset f(z)\}$ has a common face which is clearly the unique element of smallest dimension $s_f(z)$. So the definition and (8.8) follow. If $z_1 < z$ then $f(z_1) \subset N^\circ(s_f(z), K')$ and so by (8.8) $s_f(z_1) < s_f(z)$. (8.9) is then obvious. Furthermore, if $f(z) = w$ for some simplex w of K then by (8.8) $s_f(z) < w$ with equality because $w \not\subset N^\circ(w_1, K')$ for any proper face w_1 of w. So (8.10) follows. □

From condition (2) it is clear that if L is fine for f and K then any subdivision L^* of L is fine for f and K. For any L we now show that we can always obtain fineness by subdividing.

3 Proposition. *Let* $f : Y \to X$ *be a continuous map of polyhedra and let* L *and* K *be simplicial complexes with* $|L| = Y$ *and* $|K| = X$. *Assume that* L_0 *is a subcomplex of* L *and* $f_0 : L_0 \to K$ *is a simplicial map such that the restriction of* f *to* $|L_0|$ *is the p.l. map* $f_0 : |L_0| \to |K|$. *There exists* L^* *a subdivision of* L *rel* L_0 *such that* L^* *is fine for* f *and* K. *(*L^* *is a* *subdivision rel L_0 if $z \in L_0$ implies $z \in L^*$, i.e. the simplices of L_0 are not properly subdivided (see Appendix: Chapter 10).)*

Proof. For two nonempty subsets A_1 and A_2 of X the distance $d_K(A_1, A_2)$ is defined by

$$d_K(A_1, A_2) = \inf\{d_K(y_1, y_2) : y_1 \in A_1 \text{ and } y_2 \in A_2\}. \tag{8.11}$$

Notice: this is not the Hausdorff distance of (4.54).

We define the *thickness* of K to be

$$th(K) = \min_{w \in K} d_K(w, C(w, K')). \tag{8.12}$$

So by definition, for all $w \in K$ and $y \in w$ $d_K(y, y_1) < th(K)$ implies $y_1 \in N^\circ(w, K')$. By compactness $th(K)$ is positive.

Because f is uniformly continuous, we can choose $\epsilon > 0$ such that for $x_1, x_2 \in Y$

$$d_L(x_1, x_2) \leq \epsilon \Rightarrow d_K(f(x_1), f(x_2)) < th(K). \tag{8.13}$$

By Proposition 10.2 of the Appendix, there exists a subdivision L^* of L rel L_0 such that for $z^* \in L^*$

$$z^* \cap |L_0| = \emptyset \Rightarrow d_L(z^*) < \epsilon$$

$$z^* \cap |L_0| \neq \emptyset \Rightarrow d_L(z^*, L_0) < \epsilon \tag{8.14}$$

where $d_L(z^*)$ is the diameter of z^* and $d_L(z^*, L_0)$ is the maximum distance from a point of z^* to the face $z^* \cap |L_0|$. Hence, if $z^* \cap |L_0| = \emptyset$, $x \in z^*$ and $f(x) \in w \in K$ then $f(z^*) \subset N^\circ(w, K')$. If $z^* \cap |L_0|$ is the face $z_0 \in L_0$ and $w_0 = f(z_0)$ then $f(z^*) \subset N^\circ(w_0, K')$. Hence, condition (2) of Proposition 1 holds for L^*. □

Assume that $s : L \to K$ is an incidence preserving map and L_0 is a subcomplex of L on which s is a simplicial map. Let L^* be a first derived subdivision of L rel L_0 obtained by starring the simplices of $L \backslash L_0$ in descending order of dimension. A typical simplex of L^* is thus of the form $[b(z_1), \ldots, b(z_k)]$ where $z_1 < \ldots < z_k$ is a strictly increasing sequence of faces in $L \backslash L_0$ or the join of such a simplex with $z_0 \in L_0$ where $z_0 < z_1$ (see Chapter 10). For each $z \in L \backslash L_0$ choose

$$g(b(z)) \in V(s(z)) \tag{8.15}$$

For each vertex v of L_0 let $g(v) = s(v)$, a vertex of K.

4 Lemma. *The map from $V(L^*)$ to $V(K)$ defined above extends to define a simplicial map $g : L^* \to K$. Such a simplicial map is said to be* associated *with s. For every $z \in L$, the polyhedron $g(z)$ is contained in the simplex $s(z)$.*

Proof. The simplex $[b(z_1), \dots, b(z_k)]$ of L^* is carried by $z_k \in L$. Because s is incidence preserving, $z_1 < \dots < z_k$ implies

$$V(s(z_1)) \subset \dots \subset V(s(z_k)). \tag{8.16}$$

Furthermore, if $z_0 \in L_0$ with $z_0 < z_1$ then $V(s(z_0)) \subset V(s(z_1))$, as well. Hence, g maps the vertices of $[b(z_1), \dots, b(z_k)]$ (and, in the latter case the vertices of the join $z_0[b(z_1), \dots, b(z_k)]$) to vertices of $s(z_k)$ in K. Hence, by (2.20) g extends to a simplicial map. Furthermore the L^* simplices contained in $z \in L$ are all mapped to faces of $s(z)$. □

5 Proposition. *Let $f : Y \to X$ be a continuous map between polyhedra with L and K simplicial complexes such that $|L| = Y$ and $|K| = X$. Assume that L_0 is a subcomplex of L and $f_0 : L_0 \to K$ is a simplicial map such that the restriction of f to $|L_0|$ is the p.l. map $f_0 : |L_0| \to |K|$. Assume that L is fine for f and K with $s : L \to K$ an incidence preserving function approximating f such that $s = f$ on L_0 (e.g. $s = s_f$). If L^* is a first derived rel L_0 subdivision of L and $g : L^* \to K$ is an associated simplicial map then for every $z \in L$:*

$$\begin{array}{ccc} g(z) \subset & s(z) & \\ & \cap & \\ f(z) \subset & N^\circ(s(z), K'). & \end{array} \tag{8.17}$$

For every $z \in L_0$, $f = g$ on z.

Proof. This is clear from Proposition 2 and Lemma 4. □

The p.l. round off of a continuous map f on a polyhedron X is closely related to the above construction:

6 Proposition. *Let L be a simplicial complex with $X = |L|$ and let f :*

$X \to X$ be a continuous map. Let K_1 be a subdivision of L and denote by K the barycentric subdivision of K_1. Let K_1^* be a subdivision of K and let K^* denote the barycentric subdivision of K_1^*.

Assume that K_1^* is fine for f and K_1, with $s : K_1^* \to K_1$ an incidence preserving map approximating f. There is a simplicial map $g : K^* \to K$ defined on vertices by:

$$g(b(\sigma^*)) = b(s(\sigma^*)) \text{ for all } \sigma^* \in K_1^*. \tag{8.18}$$

K^* is a proper subdivision of K so that g is a simplicial dynamical system. The associated p.l. map g on X satisfies

$$d_L\left(f(x), g(x)\right) \leq 2d_L(K_1) \text{ for all } x \in X, \tag{8.19}$$

where $d_L(K_1)$ is the d_L mesh *of K_1, i.e. the maximum of the d_L diameters of the simplices of K_1. The simplicial map g and its associated p.l map is called a* p.l. roundoff *for f.*

Proof. As in Lemma 4, $\sigma_1^* < \ldots < \sigma_k^*$ implies $s(\sigma_1^*) < \ldots < s(\sigma_k^*)$ although the latter sequence need not be strictly increasing. However, it still follows that the simplex $[b(\sigma_1^*), \ldots, b(\sigma_k^*)]$ of K^* is mapped by g to a simplex with vertex set $\{b(s(\sigma_1^*)), \ldots, b(s(\sigma_k^*))\}$ in K. Hence, the simplicial map $g : K^* \to K$ is defined. Furthermore, as in (8.17), $\sigma^* \in K_1^*$ implies $g(\sigma^*) \subset s(\sigma^*)$ and $f(\sigma^*) \subset N^\circ(s(\sigma^*), K)$. It follows that for any $x \in \sigma^*$, $d_L(f(x), g(x)) \leq d_L(s(\sigma^*)) + d_L(w)$ for some $w \in K$ such that $w \cap s(\sigma^*) \neq \emptyset$. Finally, K_1^* is a subdivision of K by hypothesis and K^* is the barycentric subdivision of K_1^*. Hence, K^* is a proper subdivision of K by Lemma 2.8.

Remark: Following the proof of Proposition 3 we can choose ϵ a modulus of uniform continuity for $\mathrm{th}_L(K_1)$, i.e. as in (8.13):

$$\begin{aligned} d_L(x_1, x_2) \leq \epsilon \quad &\Rightarrow \quad d_L(f(x_1), f(x_2)) < \mathrm{th}_L(K_1), \\ \text{where } \mathrm{th}_L(K_1) \quad &= \quad \min_{\sigma \in K_1} d_L(\sigma, C(\sigma, K)). \end{aligned} \tag{8.20}$$

As in Proposition 3, K_1^* is fine for f and K_1 provided that the mesh $d_L(K_1^*)$ is at most ϵ. □

As simplicial dynamical systems, these roundoff maps have the dynamical structure that we have spent seven chapters analyzing. Because of this special origin they sometimes have some additional structure.

As in Proposition 6 let f be a continuous map on a polyhedron X with L a simplicial complex triangulating X. Let K_1 be a subdivision of L with barycentric subdivision K. We say that f is in *general position with respect to* K_1 if for $k = 0, 1, \ldots, \dim X$

$$f(|\overline{S}^k(K_1)|) \subset N^\circ(|\overline{S}^k(K_1)|, K). \tag{8.21}$$

Assume that K_1^* is a subdivision of K with barycentric subdivision K^*. Let $s : K_1^* \to K_1$ be an incidence preserving map. We say that s satisfies *Condition GP* if for every $\sigma^* \in K_1^*$ and $k = 0, \ldots, \dim X$:

$$\sigma^* \cap |\overline{S}^k(K_1)| \neq \emptyset \Rightarrow \dim s(\sigma^*) \leq k. \tag{8.22}$$

Let $g : K^* \to K$ be the simplicial map associated with s, i.e. g is defined on vertices by (8.18). Since the carrier in K_1 of $b(s(\sigma^*))$ is $s(\sigma^*)$ it follows that (8.22) is equivalent to

$$\sigma^* \cap |\overline{S}^k(K_1)| \neq \emptyset \Rightarrow g(b(\sigma^*)) \in |\overline{S}^k(K_1)|. \tag{8.23}$$

7 Proposition. *Let f be a continuous map on a polyhedron X with L a simplicial complex such that $|L| = X$. Let K_1 be a subdivision of L with barycentric subdivision K.*

(a) If f is in general position with respect to K_1 then there exists $\epsilon > 0$ such that if K_1^* is a subdivision of K with mesh, $d_L(K_1^*)$, at most ϵ then K_1^* is fine for f and K_1 and the minimal incidence preserving map $s_f : K_1^* \to K_1$ which approximates f (see Proposition 2) satisfies Condition GP.

(b) For any $\epsilon > 0$ there exists a continuous map f_1 on X such that $d_L(f(x), f_1(x)) \leq \epsilon$ for all $x \in X$ and f_1 is in general position with respect to K_1.

Proof. (a) By analogy with the definition (8.12) of thickness define for $k = 0, 1, \ldots, d = \dim X$

$$\mathrm{th}^k(K_1) = \inf_{x \in |\overline{S}^k(K_1)|} \Big(\max_{\sigma \in \overline{S}^k(K_1)} (d_L(f(x), C(\sigma, K))) \Big). \tag{8.24}$$

Notice that f is in general position with respect to K_1 iff for every $x \in |\overline{S}^k(K_1)|$ there exists $\sigma \in \overline{S}^k(K_1)$ such that $f(x) \in N^\circ(\sigma, K)$. So f is in general position iff $\mathrm{th}^k(K_1) > 0$ for $k = 0, 1, \ldots, d$. By uniform continuity we can choose $\epsilon > 0$ such that

$$(8.25) \qquad d_L(x_1, x_2) \leq \epsilon \Rightarrow d_L(f(x_1), f(x_2)) < \min_{k=0}^{d} \mathrm{th}^k(K_1).$$

Now suppose K_1^* has mesh at most ϵ and that $\sigma^* \in K_1^*$ meets $|\overline{S}^k(K_1)|$. Let $x \in \sigma^* \cap |\overline{S}^k(K_1)|$. By definition of th^k there exists $\sigma \in K_1$ with $\dim \sigma \leq k$ such that $d_L(f(x), C(\sigma, K)) \geq \mathrm{th}^k$. Because the d_L diameter of σ^* is at most ϵ, (8.25) implies $f(\sigma^*) \subset N^\circ(\sigma, K)$ which is the complement of $C(\sigma, K)$. So by Proposition 2, $s_f(\sigma^*)$ is defined and is a face of σ. Hence, K_1^* is fine for f and K_1 and s_f satisfies Condition GP.

(b) Let $z = [b(\sigma_0), \ldots, b(\sigma_n)]$ with σ_{i-1} a proper face of σ_i $(i = 1, \ldots, n)$ be a simplex of K. Clearly, $z \in C(|\overline{S}^k(K_1)|, K)$ iff $\sigma_0 \notin \overline{S}^k(K_1)$, i.e. $\dim \sigma_0 > k$. Since $\dim \sigma_{i-1} < \dim \sigma_i \leq d$ for $i = 1, \ldots, n$ it follows that $n \leq d - k$ and so we have

$$(8.26) \qquad \dim C(|\overline{S}^k(K_1)|, K) \leq d - k - 1.$$

By the p.l. approximation theorem (see, e.g. Rourke and Sanderson (1972) Section 5.4) we can assume f is a p.l. map so that $f(|\overline{S}^k(K_1)|)$ is a subpolyhedron of X with dimension $\leq k$. If X is a p.l. manifold then by the General Position Theorem for Subsets, see Rourke and Sanderson (1972) Section 5.3, there exists a p.l. homeomorphism $h : X \to X$ with $d(h(y), y) \leq \epsilon$ for all $y \in X$ and such that $h(f(|\overline{S}^k(K_1)|)) \cap C(|\overline{S}^k(K_1)|, K') = \emptyset$. Proceed inductively upward from $k = 0$ to dim X to obtain an h which works for all k. Then let $\tilde{f} = h \circ f$.

For a general polyhedron, one uses Armstrong and Zeeman's notion of intrinsic dimension (cf. Armstrong (1967)) to apply manifold tools. For $k \leq j \leq \dim X$, the intrinsic j skeleton $I^j(X)$ is triangulated by the subcomplex $I^j(K)$ of K and $I^j(X) \backslash I^{j-1}(X)$ is a j dimensional manifold. Furthermore,

$$(8.27) \qquad C(|\overline{S}^k(K_1)|, K) \cap I^j(X) = C(|\overline{S}^k(K_1)|, I^j(K))$$

which has dimension $j - k - 1$ by (8.26). One can proceed inductively upward from $j = 0$ to dim X constructing p.l. homeomorphisms h close to the identity such that

$$(8.28) \qquad h(f(|\overline{S}^k(K_1)|)) \cap C(|\overline{S}^k(K_1)|, K) \cap I^j(X) = \emptyset$$

for all $0 \leq k < j$. Compare Akin (1969) Section 8. □

Clearly, (8.21) is an open condition with respect to C^0 perturbation. Part (b) of the above proposition shows that it is a dense condition as well.

We have studied a simplicial dynamical system $g : K^* \to K$ by mapping to $X = |K| = |K^*|$ invariant subsets for the shift on K^*. Condition GP allows us to construct continuous maps in the reverse direction to the shift on K_1. For a complex K_1 triangulating X we make the dependence of the carrier map explicit, letting $q_{K_1} : X \to K_1$ be defined by

$$\sigma = q_{K_1}(x) \Leftrightarrow x \in (\sigma)^{\circ} \tag{8.29}$$

for $x \in X$ and $\sigma \in K_1$.

8 Theorem. *Let X be triangulated by K_1, a simplicial complex with barycentric subdivision K. Let K_1^* be a subdivision of K with barycentric subdivision K^*. Let $s : K_1^* \to K_1$ be an incidence preserving map satisfying condition GP, i.e. for all $\sigma^* \in K_1^*$ and $k = 0, \dots, \dim X$, $\dim s(\sigma^*) \leq k$ if σ^* meets $|\overline{S}^k(K_1)|$. Let $g : K^* \to K$ be the simplicial map associated with s, so that $g(b(\sigma^*)) = b(s(\sigma^*))$ for all $\sigma^* \in K^*$. Let $g : X \to X$ denote the associated p.l. map. Define $q_1^+ : X \to K_1^{\mathbf{Z}_+}$ and $q_1 : X_g \to K_1^{\mathbf{Z}}$ by:*

$$q_1^+(x)_i = q_{K_1}(g^i(x)) \quad for\ i \in \mathbf{Z}_+$$

$$q_1(\xi)_i = q_{K_1}(\xi_i) \quad for\ i \in \mathbf{Z}. \tag{8.30}$$

q_1^+ maps g on X to the shift s^+ on $K_1^{\mathbf{Z}_+}$ and q_1 maps s_g on X_g to the shift s in $K_1^{\mathbf{Z}}$. On the chain recurrent set $|Cg|$ the function q_1^+ is continuous and on the chain recurrent set $|Cs_g|$ q_1 is continuous.

Proof. We first observe that (8.22), or equivalently (8.23), implies for $k = 0, 1, \dots, d = \dim X$:

$$g(N(|\overline{S}^k(K_1)|, K^*)) \subset |\overline{S}^k(K_1)|. \tag{8.31}$$

For suppose $z^* \in K^*$ with $z^* \cap |\overline{S}^k(K_1)| \neq \emptyset$. Since K^* is the barycentric subdivision of K_1^* $z^* = [b(\sigma_0^*), \dots, b(\sigma_h^*)]$ with $\sigma_0^* < \dots < \sigma_h^*$ a strictly increasing sequence of simplices in K_1^*. Since z^* is contained in the open star $\operatorname{St}(\sigma_0^*)$ and z^* meets $|\overline{S}^k(K_1)|$, it follows that σ_0^* meets $|\overline{S}^k(K_1)|$ and hence so do $\sigma_1^*, \ldots, \sigma_n^*$. So by (8.22), $s(\sigma_0^*), \ldots, s(\sigma_n^*)$ are all simplices of the k skeleton $\overline{S}^k(K_1)$. Hence, the simplex spanned by $\{b(s(\sigma_0^*)), \ldots, b(s(\sigma_n^*))\}$, i.e. $g(z^*)$, is in the barycentric subdivision of $\overline{S}^k(K_1)$.

In the language of Akin (1993) Ch., this implies that each $N(|\overline{S}^k(K_1)|, K^*)$ is an *inward set* for g, i.e. a closed set mapped by g into its own interior. Let $\epsilon > 0$ satisfy

$$\epsilon < d_{K_1}(|\overline{S}^k(K_1)|, C(|\overline{S}^k(K_1)|, K^*)) \tag{8.32}$$

for k=0,1,…,d. Any finite length ϵ chain for g which enters $N((\overline{S}^k(K_1)), K^*)$ terminates at a point in the ϵ neighborhood of $|\overline{S}^k(K_1)|$. Consequently,

$$\mathcal{C}g(N(|\overline{S}^k(K_1)|, K^*)) \subset |\overline{S}^k(K_1)| \tag{8.33}$$

for $k = 0, 1, \ldots, d$.

Now let A be a basic set for g which meets $|\overline{S}^k(K_1)|$ but not $|\overline{S}^{k-1}(K_1)|$ and let $A_g = X_g \cap (A^{\mathbf{Z}})$ be the associated basic set for s_g (see Proposition 1.1). For any $x \in A$, $A \subset \mathcal{C}g(x)$ because A is a g basic set. It follows that A is contained in $|\overline{S}^k(K_1)|$ and is disjoint from $N(|\overline{S}^{k-1}(K_1)|, K^*)$. So if we define for $\sigma \in K_1$, $O_\sigma = \sigma \backslash N(\partial\sigma, K^*)$ then

$$\begin{aligned} A &\subset |\overline{S}^k(K_1)| \backslash N(|\overline{S}^{k-1}(K_1)|, K^*) \\ &= \bigcup_{\dim\sigma = k} O_\sigma. \end{aligned} \tag{8.34}$$

Observe that if $\dim\sigma_1 = \dim\sigma = k$ then $\sigma \cap O_{\sigma_1} = \emptyset$ unless $\sigma = \sigma_1$. Thus, the union in (8.34) is disjoint and $\sigma \cap A \subset O_\sigma$. Hence, for $x \in A$

$$q_K(x) = \sigma \Leftrightarrow x \in O_\sigma. \tag{8.35}$$

Hence, if $\sigma_0, \ldots, \sigma_n$ is a sequence of simplices of K then

$$\begin{aligned} &\{x \in A : q_K(g^i(x)) = \sigma_i \text{ for } i = 0, \ldots, n\} \\ &= A \cap \bigcap_{i=0}^{n} g^{-i}(O_{\sigma_i}). \end{aligned} \tag{8.36}$$

Thus, q_1^+ is continuous on A. Similarly, q_1 is continuous on A_g.

Because $|\mathcal{C}_g|$ and $|\mathcal{C}s_g|$ are finite unions of basic sets by Theorem 4.15, the proof is complete. □

We can weaken the notion of fineness by replacing condition (8.7) by

$$f(V(z)) \subset N^\circ(s(z), K') \quad \text{for all } z \in L. \tag{8.37}$$

In other words, instead of containing the entire image $f(z)$ we assume only that $N^\circ(s(z), K')$ contains the vertex values. If this is true and we define $\tilde{s}_f(z)$ to be the smallest dimensional simplex of K such that (8.27) holds then just as in Proposition 2, $\tilde{s}_f : L \to K$ is incidence preserving and as in (8.8) we have

$$f(V(z)) \subset N^\circ(w, K') \Rightarrow \tilde{s}_f(z) < w. \tag{8.38}$$

The advantage of this variation is that fineness becomes a finite condition involving examining f only on the vertices of L. The disadvantage is that we obtain direct estimates for the distance from $f(x)$ to the p.l. approximation $g(x)$ only at vertex values for x. To obtain a C^0 estimate of the distance from f to g we require additional work using uniform continuity of f, or a Lipschitz constant for f, and the mesh of L.

9. Nondegenerate Maps on Manifolds

A polyhedron X of dimension d is called a *PL manifold* of dimension d if every point $x \in X$ has some neighborhood, itself a polyhedron, which is piecewise linearly homeomorphic to a simplex of dimension d. If X itself is p.l. homeomorphic to a simplex then X is called a *PL d ball.* If X is p.l. homeomorphic to the boundary of a $d+1$ dimensional simplex then X is called a *PL d sphere.* For example, a single point is a PL 0 ball and a pair of points is a PL 0 sphere. If K is any simplicial complex triangulating a PL d manifold X then for every $z \in K$, the *link* $\text{lk}(z; K)$ triangulates either a PL ball or a PL sphere of dimension $d - \dim z - 1$, where the link of z is the subcomplex of K consisting of all simplices of K which join to z in K:

$$\text{lk}(z; K) = \{w \in K : zw \in K\}. \tag{9.1}$$

For all of this see Hudson (1969) Section 1.5.

The results about links and joins which we will require are reviewed in the Appendix: Chapter 10. We will not require the PL theory described above. We need only the following which the reader can treat as a definition: If X is a PL d manifold and K is a triangulation of X then for every $z \in K$

$$\begin{aligned}&|\text{lk}(z; K)| \text{ is homomorphic to}\\ &\text{a ball or sphere of dimension}\\ &d - \dim z - 1.\end{aligned} \tag{9.2}$$

In particular, $|\text{lk}(z; K)|$ itself is a PL manifold. Notice that for $w \in \text{lk}(z; K)$:

$$\text{lk}(w; \text{lk}(z; K)) = \text{lk}(zw; K). \tag{9.3}$$

1 Definition. *Let $f : L \to K$ be a simplicial map and let L_0 be a subcomplex of L.*

(a) A simplex σ of L is called totally degenerate rel L_0 *if $\sigma \in L \backslash L_0$ and there exists τ a proper face of σ such that $f(\tau) = f(\sigma)$ and either $\tau \in L_0$ or dim $\tau = 0$.*

(b) A simplex σ of L is called degenerate rel L_0 *if it has a totally degenerate face. In particular, if σ is degenerate rel L_0 then $\sigma \in L \backslash L_0$.*

(c) The map f is called nondegenerate off L_0 *if no simplex of L is*

degenerate rel L_0, or equivalently, if no simplex of L is totally degenerate rel L_0. □

Observe that the notions of degenerate simplex and nondegenerate map defined in Chapter 2 agree with the above definitions in the case $L_0 = \emptyset$. Also, if f is nondegenerate off L_0 and the restriction of f to L_0 is nondegenerate then no 1 simplex of L is mapped to a vertex of K and so f is nondegenerate on all of L.

Let $ND_f(L; L_0)$ denote the set of nondegenerate rel L_0 simplices of L, i.e. the set of simplices which are not degenerate rel L_0. $ND_f(L; L_0)$ is a subcomplex of L containing L_0 and all of the vertices of L. That is,

$$L_0 \cup V(L) \subset ND_f(L; L_0). \tag{9.4}$$

Our major technical result for this chapter is a procedure which allows us to approximate a simplicial dynamical system on a manifold by a nearby nondegenerate system. The detailed argument is delicate but the underlying idea is simple enough. We pause to describe the special case of a simplicial dynamical system $f : K^* \to K$ with $|K^*| = |K|$ a PL manifold of dimension 2 without boundary. Suppose that the only totally degenerate simplex σ is of dimension 1 in K^*. Now the link $\text{lk}(\sigma; K^*)$ is a 0 sphere, a pair of points. Because $f(\sigma) = z$ is a vertex, $\text{lk}(z; K)$ is a 1 sphere. Because no 2 simplex is totally degenerate the two points of $\text{lk}(\sigma; K^*)$ map to vertices of $\text{lk}(z; K)$ rather than to z itself. Now the closed star of σ in K^* is triangulated by the join $\sigma.\text{lk}(\sigma; K^*)$ which we can subdivide to $\partial\sigma.b(\sigma).\text{lk}(\sigma; K^*)$ where $b(\sigma)$ is the barycenter of σ. Currently, f maps $b(\sigma)$ to $z = f(\sigma)$. But because $\text{lk}(z; K)$ is connected we can subdivide the segment $b(\sigma).\text{lk}(\sigma; K^*)$ in such a way that we can map it instead nondegenerately to $\text{lk}(z; K^*)$ without changing f on the two endpoints in $\text{lk}(\sigma; K^*)$. Joining with the original map f on $\partial\sigma$. We obtain a subdivision of $\sigma.\text{lk}(\sigma; K^*)$ rel $\partial\sigma.\text{lk}(\sigma; K^*)$ and a new simplicial map to K agreeing with f on $\partial\sigma.\text{lk}(\sigma; K^*)$ but which is now nondegenerate.

2 Proposition. *Let X be a PL manifold of dimension d and Y be a polyhedron of dimension at most d. Let $f : L \to K$ be a simplicial map with $|L| = Y$ and $|K| = X$. Let L_0 be a subcomplex of L.*

There exists L^* a subdivision of L rel $ND_f(L; L_0)$ and a simplicial map $g : L^* \to K$ which satisfy the following properties.

(1) $g = f$ *on the subcomplex* $ND_f(L; L_0)$ *of* L^**.*

(2) g *is nondegenerate off* L_0*.*

(3) For every vertex $v \in V(L^*)\backslash V(L)$*, i.e. every new vertex, there exists* $\sigma \in L$ *on which* f *is totally degenerate* rel L_0 *and such that* $[v, v_1]$ *is a 1 simplex of* L^* *for every vertex* v_1 *of* σ*.*

Proof. We use a double induction. First, use induction on $d = \dim X$. When $d = 0$, dim $L = \dim K = 0$ and f is nondegenerate on L. Let $L^* = L$ and $g = f$. Now assume the result for PL manifolds of dimension smaller than d.

Now use induction on the number of simplices of L which are totally degenerate rel L_0. If there are none then f is nondegenerate off L_0. Again let $L^* = L$ and $g = f$.

For the inductive step let σ be a simplex of maximum dimension in the set of totally degenerate rel L_0 simplices.

As described in Appendix: Chapter 10 we can write

$$L = c(\sigma; L) \cup (\sigma.\mathrm{lk}(\sigma; L))$$

$$c(\sigma; L) = \{\tau \in L : \tau \cap \sigma \subset \partial\sigma\}. \tag{9.5}$$

Thus, $c(\sigma; L)$ is the subcomplex of L consisting of all simplices which do not have σ as a face. See (10.8) and (10.9) which also show

$$c(\sigma; L) \cap (\sigma.\mathrm{lk}(\sigma; L)) = \partial\sigma.\mathrm{lk}(\sigma; L). \tag{9.6}$$

Because σ is totally degenerate rel L_0, we have $\sigma \notin L_0$ and there exists σ_0 such that

$$\sigma_0 \in (\partial\sigma) \cap (L_0 \cup V(L))$$

$$z = f(\sigma_0) = f(\sigma). \tag{9.7}$$

That is, σ_0 is a proper face of σ which is either a vertex or a simplex of L_0. Furthermore, σ_0 and σ have a common image under f which we denote by $z \in K$. Clearly,

$$\dim z \leq \dim \sigma_0 < \dim \sigma. \tag{9.8}$$

Because $\sigma \notin L_0$ no simplex having σ as a face is in the subcomplex L_0. In fact, since σ is totally degenerate rel L_0, all of the simplices of $L\backslash c(\sigma, L)$ are degenerate rel L_0. Consequently, by (9.4)

$$L_0 \subset ND_f(L; L_0) \subset c(\sigma; L). \tag{9.9}$$

Next observe that if $\tau \in \text{lk}(\sigma; L)$ then $f(\tau) \cap z = \emptyset$. For if not there would exist a simplex $\tau_1 \in \text{lk}(\sigma; L)$ with $f(\tau_1) < z$ and hence the join $\sigma_1 = \sigma\tau_1$ satisfies $f(\sigma_1) = z$. Because $\sigma_0 < \sigma < \sigma_1$ it would follow that σ_1 is totally degenerate rel L_0. Because σ is then a proper face of σ_1, dim $\sigma < \dim \sigma_1$. This contradicts our choice of σ as a totally degenerate simplex of maximum dimension. It follows that $\tau \in \text{lk}(\sigma; L)$ implies that $f(\tau)$ joins the simplex $z = f(\sigma)$ in K. That is, f restricts to a simplicial map

$$f_0 : \text{lk}(\sigma; L) \to \text{lk}(z; K). \tag{9.10}$$

Notice that (9.8) implies

$$\begin{aligned} \dim \text{lk}(\sigma; L) &\leq \dim L - \dim \sigma - 1 \\ &< \dim K - \dim z - 1 = d - \dim z - 1. \end{aligned} \tag{9.11}$$

Because X is a PL d manifold:

$$\begin{aligned} &\text{lk}(z; K) \text{ is a PL } d - \dim z - 1 \\ &\text{manifold homeomorphic to a ball or sphere.} \end{aligned} \tag{9.12}$$

Now we are ready to construct our subdivision. First star L at the barycenter $b(\sigma)$ of σ. As in (10.11) and Proposition 10.1a of the Appendix this is a subdivision of L rel $c(\sigma; L)$. So we define the subdivision

$$L_1^* = c(\sigma; L) \cup (\partial\sigma.b(\sigma).\text{lk}(\sigma; L)). \tag{9.13}$$

We omit parentheses for the join operations because join between complexes is associative.

Let Z_1 be the subcomplex $b(\sigma).\text{lk}(\sigma; L)$, the so-called *cone* on $\text{lk}(\sigma; L)$, with subcomplex $\text{lk}(\sigma; L)$.

Now we are ready to use our manifold hypothesis. The p.l. map $f_0 : |\text{lk}(\sigma; L)| \to |\text{lk}(z; K)|$ is a continuous map to a sphere or ball of dimension $d - \dim z - 1$ from a polyhedron of smaller dimension. It follows that the map f_0 is homotopically trivial and so can be extended over the cone. That is, we can define a continuous map

$$f_1 : |Z_1| = |b(\sigma).\text{lk}(\sigma; L)| \to |\text{lk}(z; K)|$$

(9.14) $$f_1 \| \text{lk}(\sigma; L)| = f_0.$$

Now apply Proposition 8.3 to the continuous map f_1 and the subcomplex $\text{lk}(\sigma; L)$. By that result there exists Z_2 a subdivision of $b(\sigma).\text{lk}(\sigma; L) = Z_1$ rel $\text{lk}(\sigma; L)$ and a simplicial map f_2 such that:

$$f_2 : Z_2 \to \text{lk}(z; K)$$

(9.15) $$f_2 | \text{lk}(\sigma; L) = f_0.$$

We apply the induction hypothesis to the simplicial map f_2 and the subcomplex $\text{lk}(\sigma; L)$. Observe that $|\text{lk}(z; K)|$ is a PL manifold of dimension at most $d-1$. So by the induction hypothesis there exists Z_3 a subdivision of Z_2 rel $\text{lk}(\sigma; L)$ and a simplicial map f_3 such that

$$f_3 : Z_3 \to \text{lk}(z; K)$$

(9.16) $$f_3 | \text{lk}(\sigma; L) = f_0,$$

and f_3 is nondegenerate off $\text{lk}(\sigma; L)$.

Define the subdivision L_1^* by:

(9.17) $$L_1^* = c(\sigma; L) \cup (\partial\sigma.Z_3).$$

Because Z_3 is a subdivision rel $\text{lk}(\sigma; L)$ of $b(\sigma).\text{lk}(\sigma; L)$, it follows that $\partial\sigma.Z_3$ is a subdivision rel $\partial\sigma.\text{lk}(\sigma; L)$ of $\partial\sigma.b(\sigma).\text{lk}(\sigma; L)$. So by (9.6) L_1^* is a well-defined subdivision rel $c(\sigma; L)$ of L. Define the simplicial map

$$f_4 : L_1^* \to K$$

(9.18) $$f_4 | c(\sigma; L) = f | c(\sigma; L)$$

by defining for $\sigma_1 \in \partial\sigma$ and $\sigma^* \in Z_3$:

(9.19) $$f_4(\sigma_1\sigma^*) = f(\sigma_1)f_3(\sigma^*).$$

This is well defined because $f(\sigma_1)$ is a face of $f(\sigma) = z$ and $f(\sigma^*) \in \text{lk}(z; K)$. Finally, for $\sigma^* \in Z_3$ let

$$f_4(\sigma^*) = f_3(\sigma^*) \tag{9.20}$$

which corresponds to the $\sigma_1 = \emptyset$ case of (9.19).

We now show that if $\tilde{\sigma} \in \partial\sigma.Z_3$ is totally degenerate rel L_0 for f_4 then $\tilde{\sigma} \in \partial\sigma.\text{lk}(\sigma; L)$ and so $\tilde{\sigma} \in c(\sigma; L)$. Suppose $\tilde{\sigma} = \sigma_1\sigma^*$ with $\sigma^* \in Z_3$ and $\sigma_1 \in \partial\sigma$ or $\sigma_1 = \emptyset$. There must exist $\tilde{\tau}$ a proper face of $\tilde{\sigma}$ with $f_4(\tilde{\tau}) = f_4(\tilde{\sigma})$ and either $\tilde{\tau} \in L_0 \subset c(\sigma; K)$ or dim $\tilde{\tau} = 0$. We can write

$$\tilde{\tau} = \tau_1\tau^* \text{ with}$$

$$\tau_1 = \tilde{\tau} \cap \sigma_1 \text{ and } \tau^* = \tilde{\tau} \cap \sigma^*. \tag{9.21}$$

By (9.19) we have

$$f(\tau_1) = f(\sigma_1) < z$$

$$f_3(\tau^*) = f_3(\sigma^*) \in \text{lk}(z; K). \tag{9.22}$$

In particular, τ^* is not empty. Since τ^* is a face of $\tilde{\tau}$, either $\tau^* \in L_0$ or dim $\tau^* = 0$. Since $L_0 \subset c(\sigma; L)$ we have either dim $\tau^* = 0$ or

$$\tau^* \in c(\sigma; L) \cap Z_3 \subset \text{lk}(\sigma; L). \tag{9.23}$$

But by construction f_3 is nondegenerate off $\text{lk}(\sigma; L)$. In particular, σ^* is not totally degenerate rel $\text{lk}(\sigma; L)$. This implies τ^* is not a proper face of σ^*. So $\tau^* = \sigma^*$. Because $\tilde{\tau}$ is a proper face of $\tilde{\sigma}$ it must be true that τ_1 is a proper face of σ_1. In particular, by (9.22) τ_1 and σ_1 are nonempty. But then $f(\sigma_1)$ and $f(\tau^*)$ joinable imply that dim $f(\tilde{\tau}) \neq 0$ and so dim $\tilde{\tau} \neq 0$, which implies $\tilde{\tau} \in L_0$. Hence, the face τ^* is in L_0 and $\tau^* = \sigma^*$ implies $\sigma^* \in L_0$. Thus, $\sigma^* \in L_0 \cap Z_3 \subset \text{lk}(\sigma; L)$ and $\tilde{\sigma} \in \partial\sigma.\text{lk}(\sigma; L)$ as required.

It follows that the set of simplices of K_1^* which are totally degenerate rel L_0 for f_4 are exactly the simplices of $c(\sigma; L)$ which are totally degenerate rel L_0 for f. This is exactly the original set of simplices of L which are totally degenerate rel L_0 for f with the exclusion of σ. By (9.9) we have

$$ND_f(L; L_0) \subset ND_{f_4}(L_1^*; L_0). \tag{9.24}$$

Because $f_4 : L_1^* \to K$ has fewer totally degenerate simplices rel L_0 then $f : L \to K$ did, we can apply the second induction hypothesis to f_4. We obtain L^* a subdivision of L_1^* rel $ND_{f_4}(L_1^*; L_0)$ and a simplicial map

$g : L^* \to K$ which satisfies conditions (1)-(3) for f_4. We must verify conditions (1)-(3) for f.

First, L_1^* agrees with L^* on ND_{f_4} and so by (9.24) on ND_f. L_1^* agrees with L on $c(\sigma; L)$ and so by (9.9) on ND_f. Thus, L^* is a subdivision rel ND_f of L. Similarly, $g = f_4$ on ND_{f_4} and $f_4 = f$ on $c(\sigma; L)$ imply together $g = f$ on ND_f. By construction g is nondegenerate off L_0.

Finally, let $v \in V(L^*)\backslash V(L)$. If $v \in V(L^*)\backslash V(L_1^*)$ then by construction of g there exists $\tilde{\sigma}$ totally degenerate rel L_0 for f_4 such that $[v, v'] \in L^*$ for all $v' \in V(\tilde{\sigma})$. But $\tilde{\sigma}$ is then contained in $c(\sigma; L)$ and is a totally degenerate simplex for f. Hence, v satisfies condition (3). What remains is the case $v \in V(L_1^*)\backslash V(L) = V(Z_3)\backslash V(\text{lk}(\sigma; L))$. Use the original totally degenerate simplex σ. If v' is any vertex of σ then dim $\sigma > 0$ implies $v' \in \partial\sigma$ and so $[v', v] \in L_1^*$. As they are vertices, neither v nor v' is totally degenerate rel L_0 for f_4. Since $f_4(v') \in z$ and $f_4(v) \in \text{lk}(z; K)$ it follows that the one simplex $[v', v]$ is nondegenerate for f_4. That is, $[v', v] \in ND_{f_4}(L_1^*; L)$. Because L^* is a subdivision of L_1^* rel ND_{f_4} it follows that $[v', v] \in L^*$ proving condition (3) in this case as well.

Remark: The peculiar condition (3) is a device to control throughout a long induction the degree of approximation to the original function. □

3 Theorem. *Let X be a PL manifold triangulated by a simplicial complex L. Assume that $f : K^* \to K$ is a simplicial dynamical system with K a subdivision of L. Let $ND(K^*)$ be the subcomplex of K^* on the simplices of which f is nondegenerate.*

There exists K^{**} a subdivision rel $ND(K^*)$ of K^* and a nondegenerate simplicial map $g : K^{**} \to K$ such that the p.l. maps f and g on X satisfy:

$$g(x) = f(x) \text{ for all } x \in |ND(K^*)| \tag{9.25}$$

and

$$d_L(g(x), f(x)) \leq 4d_L(K) \text{ for all } x \in X. \tag{9.26}$$

where $d_L(K)$ is the mesh of K measured using the metric d_L.

Proof. Apply Proposition 2 with $L_0 = \emptyset$. Equation (9.25) is obvious because $g = f$ on the subcomplex $ND(K^*)$. If $x \in X$, let z^{**} and z_1^* be the carriers of x in K^{**} and K^* respectively so that $z^{**} \subset z_1^*$. Let v_1 be a vertex of z^{**}. Either $v_1 \in V(K^*) \subset ND(K^*)$ in which case define $v_2 = v_1$. Or else $v_1 \in V(K^{**})\backslash V(K^*)$ and so by condition (3) of Proposition 2 there exists $v_2 \in V(K^*) \subset ND(K^*)$ such that $[v_1, v_2]$ is a 1 simplex of K^{**} and so is contained in some simplex z_2^* of K^*. By $v_2 \in ND(K^*)$ we have $f(v_2) = g(v_2)$. The segment $g([x, v_1]) \subset g(z^{**}) \in K$ and $g([v_1, v_2]) \in K$. The segment $f([v_1, v_2])$ is contained in $f(z_2^*) \in K$ and $f([x, v_1])$ is a subset of $f(z_1^*) \in K$. By the triangle inequality $d_L(g(x), f(x)) \leq 4d_L(K)$.

Remark. By Lemma 2.8d, K^{**} is a proper subdivision of K since K^* is, and so $g : K^{**} \to K$ is a simplicial dynamical system. □

It is a classical approximation result of PL topology that a continuous map on a PL manifold can be uniformly approximated arbitrarily closely by a nondegenerate p.l. map. The delicacy arose because we wished to avoid subdividing the range complex.

4 Lemma. *Let L_0 be a subcomplex of a simplicial complex L. If L_0^* is a subdivision of L_0 then there exists L^* a subdivision of L such that L_0^* is a subcomplex of L^*.*

Proof. Proceed by induction on dim $L = d$. The result is trivial when $d = 0$. The inductive hypothesis allows us to assume that L_0 contains the $d - 1$ skeleton of L. Now for $\sigma \in L$ with dim $\sigma = d$. Let Z_σ be the subdivision of $\partial\sigma$ induced by L_0^*. Then $b(\sigma).Z_\sigma$ is a subdivision of σ which agrees with Z_σ on the boundary. Thus, we extend L_0^* over the interior of each d simplex by coning from the barycenter. □

5 Proposition. *Let K^* be a subdivision of a simplicial complex K. There exists K^{**} a proper subdivision of K^* and a nondegenerate simplicial map $g : K^{**} \to K$ such that the p.l. map g on $|K|$ is subordinate to the identity 1_K, i.e. for each $z \in K$, $g(z) = z$.*

Proof. Let K_1^* be the barycentric subdivision of K^*. Now proceed again by induction on dim $K = d$. As usual the result is trivial when $d = 0$.

Let $Y = |\overline{S}^{d-1}(K)|$. For any subdivision $\tilde{K}$ of K let $\tilde{K} \wedge Y$ denote the subcomplex of $\tilde{K}$ triangulating Y.

By induction hypothesis and Lemma 4 there exists K_2^* a subdivision of K_1^* and a simplicial map $f_0 : K_2^* \wedge Y \to K \wedge Y$ such that $f_0(z) = z$ for all

$z \in K$ with dim $z < d$.

For $z \in K$ with dim $z = d$ let $K_2^* \wedge z$ and $K_2^* \wedge \partial z$ denote the subcomplexes induced on z and ∂z respectively by K_2^*. By construction, f_0 restricts to a nondegenerate simplicial map of $K_2^* \wedge \partial z$ to z. Fix some vertex v of z and map the vertices of $V(K_2^* \wedge z) \backslash V(K_2^* \wedge \partial z)$ to v. This extends f_0 to a simplicial map

$$f_z : K_2^* \wedge z \to z$$

(9.27) $$f_z = f_0 \text{ on } K_2^* \wedge \partial z.$$

Of course, f_z is usually very degenerate. Apply Proposition 2 with $L = K_2^* \wedge z$, $f = f_z$ and $L_0 = K_2^* \wedge \partial z$. Because z is a PL manifold, we obtain Z_z a subdivision rel $K_2^* \wedge \partial z$ of $K_2^* \wedge z$ and a simplicial map

$$g_z : Z_z \to z, \text{ such that}$$

(9.28) $$g_z = f_z = f_0 \text{ on } K_2^* \wedge \partial z,$$

with g_z nondegenerate off $K_2^* \wedge \partial z$. Because f_0 is nondegenerate on $K_2^* \wedge \partial z$ it follows that g_z is nondegenerate.

Define:

(9.29) $$K^{**} = \Big(\bigcup_{\dim z = d} Z_z \Big) \cup (K_2^* \wedge Y)$$

and $f : K^{**} \to K$ by $f = f_0$ on $K_2^* \wedge Y$, $f = g_z$ on Z_z for each $z \in S^d(K)$.

Because K^{**} is a subdivision of K_1^* it is a proper subdivision of K^*. Clearly, $f : K^{**} \to K$ is nondegenerate and subordinate to 1_K. □

In particular, with $K^* = K$ there is a proper subdivision K^{**} of K and a nondegenerate simplicial map $f : K^{**} \to K$ subordinate to 1_K. The simplest example I know of this is the lunar subdivision constructed in Chapter 10 of the Appendix.

6 Theorem. *Let X be a PL manifold of dimension d, triangulated by a simplicial complex L. Let f be a chain transitive continuous map on X, i.e. $Cf = X \times X$. Let K be a subdivision of L.*

There exists a proper subdivision K^* of K and a nondegenerate simplicial map $g : K^* \to K$ whose associated p.l. map g on X is topologically transitive and which satisfies

(9.30) $$d_L(g(x), f(x)) \leq 9 d_L(K) \text{ for all } x \in X$$

when $d_L(K)$ is the mesh of K measured using the metric d_L.

Furthermore, with $S^d(K^*) = \{z^* \in K^* : \dim z^* = d\}$ the subshifts $s^+_{G^*}$ on $S^d(K^*)^+_{G^*}$ and s_{G^*} on $S^d(K^*)_{G^*}$ are transitive subshifts. In addition, $h^+ : S^d(K^*)^+_{G^*} \to X$ and $h : S^d(K^*)_{G^*} \to X_g$ are almost homeomorphisms mapping $s^+_{G^*}$ to g and s_{G^*} to s_g, respectively.

If, addition, X is connected then the subshifts $s^+_{G^*}$ on $S^d(K^*)^+_{G^*}$ and s_{G^*} on $S^d(K^*)_{G^*}$ are mixing and the p.l map g on X is topologically mixing.

Proof. Let S denote $S^d(K^*)$ and S^* denote $S^d(K^*)$. On the finite set S we define the relation $S_f \subset S \times S$ by

$$(z_1, z_2) \in S_f \Leftrightarrow z_1 \cap f^{-1}(N^\circ(z_2, K')) \neq \emptyset \tag{9.31}$$

where K' is the barycentric subdivision of K. Since $f^{-1}(N^\circ)$ is open and $(z_1)^\circ$ is dense in z_1 we have

$$(z_1, z_2) \in S_f \Leftrightarrow (z_1)^\circ \cap f^{-1}(N^\circ(z_2, K')) \neq \emptyset. \tag{9.32}$$

Next we prove that S_f has a single basic set consisting of all of S, i.e.

$$\mathcal{O}S_f = S \times S. \tag{9.33}$$

To prove this let $z, \tilde{z}$ and choose $x \in z$ and $\tilde{x} \in \tilde{z}$. Let $\delta_0 > 0$ be smaller than the thickness of S, i.e. $\min_{z \in S} d_L(z, C(z, K'))$ (see (8.20)). Because f is chain transitive there exists a δ_0 chain $\{x_0, x_1, \ldots, x_n\}$ for f with $n \geq 1$, $x_0 = x$ and $x_n = \tilde{x}$. Since X is covered by the d simplices we can choose $\{z_0, z_1, \ldots, z_n\}$ in S such that $z_0 = z$, $z_n = \tilde{z}$ and $x_i \in z_i$ for $i = 0, \ldots, n$. Because $d(f(x_i), x_{i+1}) \leq \delta_0$ for $i = 0, \ldots, n-1$, we have $x_i \in z_i \cap f^{-1}(N^\circ(z_{i+1}, K'))$ for $i = 0, \ldots, n-1$. That is, $\{z_0, \ldots, z_n\}$ is an S_f chain from z to $\tilde{z}$.

For each pair $(z_1, z_2) \in S_f$ apply (9.32) to choose $a(z_1, z_2) \in X$ with

$$a(z_1, z_2) \in (z_1)^\circ \cap f^{-1}(N^\circ(z_2, K')). \tag{9.34}$$

Furthermore, we can choose the points so that for $(z_1, z_2), (\tilde{z}_1, \tilde{z}_2) \in S_f$

$$(z_1, z_2) \neq (\tilde{z}_1, \tilde{z}_2) \Rightarrow a(z_1, z_2) \neq a(\tilde{z}_1, \tilde{z}_2). \tag{9.35}$$

This is automatically true when the dimension $d = 0$. When $d > 0$ the open sets $(z_1)^\circ \cap f^{-1}(N^\circ(z_2, K'))$ are infinite and so we can make distinct choices for the finite set S_f.

With $V_\delta(x) = \{x_1 : d_L(x, x_1) < \delta\}$ we can choose $\delta_1 > 0$ so that

$$V_{\delta_1}(a(z_1, z_2)) \subset (z_1)^\circ \cap f^{-1}(N^\circ(z_2, K')) \tag{9.36}$$

for all $(z_1, z_2) \in S_f$, and, also,

$$V_{\delta_1}(a(z_1, z_2)) \cap V_{\delta_1}(a(\tilde{z}_1, \tilde{z}_2)) = \emptyset \tag{9.37}$$

if (z_1, z_2) and $(\tilde{z}_1, \tilde{z}_2)$ are distinct elements of S_f.

Let K_1^* be a proper subdivision of K with $d_L(K_1^*) < \delta_1/2$. Choose for each $(z_1, z_2) \in S_f$ a simplex $\sigma(z_1, z_2)$ of K_1^* such that

$$\begin{aligned} a(z_1, z_2) &\in \sigma(z_1, z_2) \\ \dim \sigma(z_1, z_2) &= d. \end{aligned} \tag{9.38}$$

Let $N(z_1, z_2)$ and $N^\circ(z_1, z_2)$ denote $N(\sigma(z_1, z_2), K_1^{*\prime})$ and $N^\circ(\sigma(z_1, z_2), K_1^{*\prime})$ respectively. Because the mesh of K_1^* is smaller than $\delta_1/2$, (9.38) and (9.36) imply

$$N(z_1, z_2) \subset (z_1)^\circ \cap f^{-1}(N^\circ(z_2, K')) \tag{9.39}$$

for all $(z_1, z_2) \in S_f$. Similarly, (9.37) implies

$$N(z_1, z_2) \cap N(\tilde{z}_1, \tilde{z}_2) = \emptyset \tag{9.40}$$

for (z_1, z_2), $(\tilde{z}_1, \tilde{z}_2)$ distinct elements of S_f.

Now define a continuous map f_1 on X so that

$$f_1 = f \quad \text{on } X \setminus \bigcup_{(z_1, z_2) \in S_f} N^\circ(z_1, z_2) \tag{9.41}$$

and for each $(z_1, z_2) \in S_f$, f_1 is a linear isomorphism from the d simplex $\sigma(z_1, z_2)$ to the d simplex z_2. Use an arbitrary bijection of the vertex sets.

Because X is a PL manifold, the regular neighborhood $N(z_2, K')$ (see Cohen (1969)) is a PL ball. So far f_1 is defined and maps the subsets $N(z_1, z_2) \setminus N^\circ(z_1, z_2)$ and $\sigma(z_1, z_2)$ into $N(z_2, K')$ by (9.39) and (9.41). So we can use, e.g. the Tietze Extension Theorem, to define f_1 on $N(z_1, z_2)$ so that

$$(9.42) \qquad f_1(N(z_1, z_2)) \subset N(z_2, K').$$

By (9.40) the definitions on $N(z_1, z_2)$ and $N(\tilde{z}_1, \tilde{z}_2)$ for distinct $(z_1, z_2), (\tilde{z}_1, \tilde{z}_2) \in S_f$ do not interfere.

By (9.42) and (9.39) we have $f(x), f_1(x) \in N(z_2, K')$ for all $x \in N(z_1, z_2)$. Together with (9.41) this implies

$$(9.43) \qquad d_L(f(x), f_1(x)) \leq 3d_L(K)$$

for all $x \in X$.

Now apply the approximation theorem Proposition 8.5 to $f_1 : |K_1^*| \to |K|$ with L_0 the subcomplex consisting of $\{\sigma(z_1, z_2) : (z_1, z_2) \in S_f\}$ together with all of its faces. We obtain a subdivision K_2^* rel L_0 and a simplicial map $f_2 : K_2^* \to K$ such that for $(z_1, z_2) \in S_f$

$$(9.44) \qquad f_2 = f_1 \text{ on } \sigma(z_1, z_2)$$

and by (8.17) the p.l. map f_2 on X satisfies

$$(9.45) \qquad d_L(f_1(x), f_2(x)) \leq 2d_L(K).$$

Now we apply Proposition 2 to the simplicial map $f_2 : K_2^* \to K$ and the subcomplex L_0. We obtain K^* a subdivision rel L_0 of K_2^* and a simplicial map $g : K^* \to K$. For each $(z_1, z_2) \in S_f$

$$(9.46) \qquad g = f_2 = f_1 \text{ on } \sigma(z_1, z_2).$$

In particular, g is nondegenerate on L_0. As g is nondegenerate off L_0 it follows that g is nondegenerate. By Condition (3) of Proposition 2 together with $g = f_2$ on $V(K_2^*)$ it follows as in the proof of Theorem 3 that

$$(9.47) \qquad d_L(g(x), f_2(x)) \leq 4d_L(K)$$

for all $x \in X$. So (9.30) follows from (9.43) (9.45) and (9.47). Because K_1^* is a proper subdivision of K and K^* is a subdivision of K_1^*, K^* is proper for K (see Lemma 2.8). Thus, $g : K^* \to K$ is a nondegenerate simplicial dynamical system, with associated relations G^* on K^* and G on K.

For every $(z_1, z_2) \in S_f$ we have $\sigma(z_1, z_2) \in K^*$ with $\sigma(z_1, z_2) \subset z_1$ and $g(\sigma(z_1, z_2)) = z_2$. This says that $(z_1, z_2) \in G$. So we have

$$(9.48) \qquad S_f \subset G.$$

So by (9.33) the entire set S of dimension d simplices of K forms the single d skeleton basic set for G. Because g is nondegenerate (4.10) implies that the entire set S^* of dimension d simplices of K^* is the associated G^* basic set. Hence, $s^+_{G^*}$ on $S^{*+}_{G^*}$ and s_{G^*} on $S^*_{G^*}$ are transitive subshifts. By Theorem 4.17, $h^+ : S^{*+}_{G^*} \to X$ and $h : S^*_{G^*} \to X_g$ are almost homeomorphisms mapping the shifts to g and s_g respectively. Consequently these maps are topologically transitive. In fact the whole space X is a single basic set image for g.

Now assume that X is connected. Because f is chain transitive it follows that it is chain mixing (see Akin (1993) Exercise 8.22) which implies that given $\delta_0 > 0$ there exists a positive integer N such that every pair of points in X can be joined by a δ_0 chain of length N. This allows us to sharpen (9.33) to get

(9.49) $$(S_f)^N = S \times S$$

by adapting the proof of (9.33) in the obvious way. From (9.48) it follows that

(9.50) $$G^N = S \times S.$$

If $z_1^*, z_2^* \in S^*$ then $(g(z_1^*), j(z_2^*)) \in S \times S = G^N$ and so $(z_1^*, z_2^*) \in (G^*)^{N+1}$. Hence

(9.51) $$(G^*)^{N+1} = S^* \times S^*.$$

This implies that the transitive subshifts $s^+_{G^*}$ and s_{G^*} are aperiodic and so are mixing subshifts. As continuous maps they are topologically mixing and so the factor maps g and s_g are topologically mixing. □

A continuous map f on a compact metric space X is *chain transitive* if $\mathcal{C}f = X \times X$. It is *topologically transitive* when $\mathcal{N}f = X \times X$ where

(9.52) $$\mathcal{N}f = \overline{\cup_{n=1}^{\infty} f^n} \subset X \times X.$$

By Akin (1993) Theorem 4.12 this is equivalent to the definition given in Chapter 1 that the set of transitive points is nonempty. As described in Chapter 1, f is *topologically transitive* when for every pair of nonempty open sets U and V,

(9.53) $$U \cap f^{-k}(V) \neq \emptyset$$

for infinitely many positive integers k, while f is *topologically mixing* if (9.53) holds for all k sufficiently large. The map f is called *weak mixing* when $f \times f$ is *topologically transitive* on $X \times X$ (see eg. Auslander (1988) Chapter 9). Topological mixing implies weak mixing and weak mixing implies topological transitivity.

Let $C(X;X)$ denote the space of continuous maps on X with the uniform topology. That is, for $f_1, f_2 \in C(X;X)$ we define

$$d(f_1, f_2) = \sup_{x \in X} d(f_1(x), f_2(x)). \tag{9.54}$$

$C(X;X)$ is a complete metric space and so is a Baire space.

From Theorem 6 we obtain

7 Corollary. *Let X be a PL manifold. The set $\{f \in C(X;X) : \mathcal{C}f = X \times X\}$ of chain transitive maps is a closed subset of $C(X;X)$ containing $\{f \in C(X;X) : \mathcal{N}f = X \times X\}$, the set topologically transitive maps, as a dense, G_δ subset. If X is a connected PL manifold then the set of chain transitive maps is nonempty and contains $\{f \in C(X;X) : \mathcal{N}(f \times f) = (X \times X) \times (X \times X)\}$, the set of weak mixing maps, as a dense, G_δ subset.*

Proof. Choose a triangulation L for X to obtain a metric d_L so that X is a compact metric space and $C(X;X)$ is a complete metric space. Following Akin (1993) Chapter 7 we put the Hausdorff metric on the space $C(X \times X)$ of closed subsets of $X \times X$. By Akin (1993) Theorem 7.24, the maps $\mathcal{C} : C(X;X) \to C(X \times X)$ and $\mathcal{N} : C(X;X) \to C(X \times X)$ defined by $f \mapsto \mathcal{C}f$ and $\mathcal{N}f$, respectively, are upper semicontinuous and lower semicontinuous, respectively. So for any open subset U of $X \times X$, $\{f : \mathcal{C}f \subset U\}$ and $\{f : \mathcal{N}f \cap U \neq \emptyset\}$ are open subsets of $C(X;X)$. By letting U vary over all proper open subsets of $X \times X$ and taking the union of the $\{f : \mathcal{C}f \subset U\}$'s we see that the complement of $\{f : \mathcal{C}f = X \times X\}$ is open. By letting U vary over a countable base of nonempty open subsets and intersecting the $\{f : \mathcal{N}f \cap U \neq \emptyset\}$'s we see that $\{f : \mathcal{N}f = X \times X\}$ is G_δ. By Theorem 6, the set of topologically transitive maps is dense in the set of chain transitive maps when X is a PL manifold.

When X is a connected metric space, the identity map 1_X is chain transitive (cf. Akin (1993) Exercise 1.9b). Since $f \mapsto f \times f$ defines a continuous map from $C(X;X)$ to $C(X \times X; X \times X)$ we see that $\{f : \mathcal{N}(f \times f) = (X \times X) \times (X \times X)\}$ is a G_δ subset of $\{f : \mathcal{C}f = X\}$. By Theorem 6, the set of topologically mixing maps is dense in the set of chain transitive maps when X is a connected PL manifold. □

Because a simplicial dynamical system is never injective when the dimension of the space is positive, this approach leaves open the question whether every chain transitive homeomorphism on a manifold can be uniformly approximated by a topologically transitive homeomorphism.

Following Oxtoby (1971) Chapter 18 one can show that on a connected manifold with a fixed Lebesgue-type measure μ, the set of weak mixing homeomorphisms f such that $f_*\mu = \mu$ is dense in, and so is a dense G_δ subset of, the set of all homeomorphisms preserving the measure. By the Poincare Recurrence Theorem if a continuous map on a compact connected metric space X preserves a measure μ with $|\mu| = X$ then f is a *central map*, i.e. $1_X \subset \mathcal{N}f$. From this it follows that $X \times X = \mathcal{C}1_X \subset \mathcal{C}(\mathcal{N}f) = \mathcal{C}f$ by Akin (1993) Proposition 1.11c. Hence, such a map is always chain transitive. However, there are plenty of chain transitive homeomorphisms which are not central.

10. Appendix: Stellar and Lunar Subdivisions

Let K be a simplicial complex and K_0 be a subcomplex of K. K^* is called a *subdivision rel* K_0 if it is a subdivision of K and $K_0 \subset K^*$. That is, the simplices of K_0 remain intact and are not further subdivided. An equivalent statement is

$$V(K^*) \cap |K_0| = V(K_0), \tag{10.1}$$

the only K^* vertices contained in the subpolyhedron $|K_0|$ are those of K_0.

To construct the stellar subdivisions we review from Hudson (1968) Section 1.2 and Cohen (1969) some results about joins and links.

If σ and τ are simplices in some vector space we call σ and τ *joinable* if there is a simplex z such that

$$V(\sigma) \cap V(\tau) = \emptyset$$

$$V(\sigma) \cup V(\tau) = V(z). \tag{10.2}$$

That is, the two vertex sets are disjoint and their union is an independent set with span z. We then write

$$z = \sigma\tau = \tau\sigma. \tag{10.3}$$

As a convention we write

$$\sigma = \emptyset\sigma = \sigma\emptyset. \tag{10.4}$$

Simplicial complexes K and L in a vector space are called *joinable* if $\sigma \in K$ and $\tau \in L$ are all joinable and $\sigma_1, \sigma_2 \in K$; $\tau_1, \tau_2 \in L$ imply

$$\sigma_1\tau_1 \cap \sigma_2\tau_2 = (\sigma_1 \cap \sigma_2)(\tau_1 \cap \tau_2). \tag{10.5}$$

In that case, we write

$$K.L = L.K = K \cup L \cup \{\sigma\tau : \sigma \in K, \tau \in L\}, \tag{10.6}$$

which is then a simplicial complex with subcomplexes K and L. Let K^* be a subdivision of K rel K_0 and L^* a subdivision of L rel L_0. It is easy to check that if K and L are joinable then K^* and L^* are joinable and $K^*.L^*$ is a subdivision of $K.L$ rel $K_0.L_0$.

As with (10.4) we use the convention

$$\emptyset.K = K.\emptyset = K. \tag{10.7}$$

If σ is a simplex of a complex K and the vertex set $V(z)$ is partitioned into two nonempty sets I, $\tilde{I}$ then the faces $\sigma_I = [I]$ and $\sigma_{\tilde{I}} = [\tilde{I}]$ are joinable with $z = \sigma_I \sigma_{\tilde{I}}$. In this case we write $\sigma_I \sigma_{\tilde{I}} \in K$. So $\sigma\tau \in K$ for $\sigma, \tau \in K$ means σ and τ are joinable and the join is a simplex of K.

For $z \in K$ define the subcomplexes of K:

$$\begin{aligned} \mathrm{lk}(z; K) &= \{\sigma \in K : \sigma z \in K\} \\ c(z; K) &= \{\sigma \in K : \sigma \cap z \subset \partial z\}. \end{aligned} \tag{10.8}$$

The first of these is called the *link* of z in K.

$$\begin{aligned} K &= c(z; K) \cup (z.\mathrm{lk}(z; K)) \\ c(z; K) \cap (z.\mathrm{lk}(z; K)) &= \partial z.\mathrm{lk}(z; K), \end{aligned} \tag{10.9}$$

where the z and ∂z joined to $\mathrm{lk}(z; K)$ refer to the complexes consisting of all of the faces, resp. the proper faces, of the simplex z.

$$\begin{aligned} |c(z; K)| &= |K| \backslash \mathrm{St}(z; K) \\ |z.\mathrm{lk}(z; K)| &= \overline{\mathrm{St}(z; K)} \end{aligned} \tag{10.10}$$

where $\mathrm{St}(z; K)$ is the open star of z, i.e. the union of all $(\sigma)^\circ$ such that $z < \sigma$.

Now suppose that L is a subdivision of z rel ∂z. We can then define a subdivision of K rel $c(z; K)$ by:

$$c(z; K) \cup (L.\mathrm{lk}(z; K)). \tag{10.11}$$

That is, the new simplices, those in $L.\mathrm{lk}(z; K)$ are of the form $z^*\tau$ for $z^* \in L$ and $\tau \in \mathrm{lk}(z; K)$.

1 Proposition. *Let* $\sigma = [v_0, \ldots, v_n]$ *be an* n *dimensional simplex. Let the* barycenter *of* σ, $b(\sigma)$, *be defined by*

$$b(\sigma) = \frac{1}{n+1} \sum_{i=0}^{n} v_i. \tag{10.12}$$

Let the ϵ reflection of v_i in σ, $r(\sigma)(v_i)$, *be defined by*

$$r(\sigma)(v_i) = b(\sigma) + \epsilon(b(\sigma) - v_i) = (1+\epsilon)b(\sigma) - \epsilon v_i, \tag{10.13}$$

for $i = 0, 1, \ldots, n$.

Here we assume $1 > n\epsilon > 0$ *so that* $r(\sigma)(v_i) \in (\sigma)^\circ$.

(a) Let $b \in (\sigma)^\circ$. *The zero simplex* b *and the boundary complex* $\partial\sigma$ *are joinable. The join* $b.\partial\sigma$ *is a subdivision of* σ *rel* $\partial\sigma$.

(b) If I *is a subset of* $\{0, \ldots, n\}$ *let* σ_I *denote the span of* $\{v_j : j \in I\}$ *and let* $\tilde{I}$ *be the complement* $\{0, \ldots, n\} \backslash I$ *so that* $\sigma = \sigma_I \sigma_{\tilde{I}}$. *If* I *is nonempty then* $\{r(\sigma)(v_j) : j \in I\}$ *is an independent subset of* $(\sigma)^\circ$. *Denote the simplex it spans by* $r(\sigma)_I$. *The simplices* $r(\sigma)_I$ *and* $\sigma_{\tilde{I}}$ *are joinable and we denote by* $r(\sigma)_I.\sigma_{\tilde{I}}$ *the complex consisting of* $r(\sigma)_I\sigma_{\tilde{I}}$ *and its faces. Define*

$$L(\sigma) = \cup_I r(\sigma)_I.\sigma_{\tilde{I}} \tag{10.14}$$

with the union over all nonempty subsets I *of* $\{0, \ldots, n\}$. $L(\sigma)$ *is a simplicial complex which is a subdivision of* σ *rel* $\partial\sigma$.

Proof: (a) Let $b = \sum_{i=0}^n b_i v_i$ with $b_i > 0$ for all i. For $x \in \sigma$ with $x = \sum_{i=0}^n x_i v_i$ let $i_0 \in \{0, \ldots, n\}$ and write

$$x = y_b b + \sum_{i=0, i \neq i_0}^{n} y_i v_i \tag{10.15}$$

By uniqueness of the barycentric coordinates:

$$x_i = \begin{cases} y_b b_i + y_i & i \neq i_0 \\ y_b b_{i_0} & i = i_0 \end{cases} \tag{10.16}$$

From which we obtain

$$\begin{aligned} y_b &= x_{i_0}/b_{i_0} \\ y_i &= x_i - x_{i_0} b_i / b_{i_0} \end{aligned} \tag{10.17}$$

This includes the definition $y_{i_0} = 0$. In order that $y_i \geq 0$ for all i, it must be that x_{i_0}/b_{i_0} is the minimum of the set $\{x_i/b_i\}$. Then by (10.17):

$$\begin{aligned} y_b &= \min_{i=0}^{n}\{x_i/b_i\} \\ y_i &= x_i - y_b b_i \end{aligned} \tag{10.18}$$

uniquely determines the coefficients of x in the complex $b.\partial\sigma$. Observe that x is in the boundary iff $y_b = 0$.

(b) Now with $x = \sum_{i=0}^{n} x_i v_i$ and I a nonempty subset of $\{0, \ldots, n\}$ assume

$$\begin{aligned} x &= \sum_{i \in I} y_i r_i + \sum_{i \in \tilde{I}} y_i v_i \\ &= (1+\epsilon) y_I b(\sigma) - \epsilon \sum_{i \in I} y_i v_i + \sum_{i \in \tilde{I}} y_i v_i. \end{aligned} \tag{10.19}$$

where we write r_i for $r(\sigma)(v_i)$ and y_I for $\sum_{i \in I} y_i$. So we have

$$x_i = \frac{1+\epsilon}{n+1} y_I + \begin{cases} -\epsilon y_i & i \in I \\ y_i & i \in \tilde{I}. \end{cases} \tag{10.20}$$

Writing $\#I$ for the cardinality of I we have, by summing over $i \in I$

$$x_I = \frac{(1+\epsilon)\#I - (n+1)\epsilon}{n+1} y_I \tag{10.21}$$

and so

$$x_i - \frac{(1+\epsilon)}{(1+\epsilon)\#I - (n+1)\epsilon} x_I = \begin{cases} -\epsilon y_i & i \in I \\ y_i & i \in \tilde{I}. \end{cases} \tag{10.22}$$

Observe that $I \neq \emptyset$ implies

$$(1+\epsilon)\#I - (n+1)\epsilon \geq (1+\epsilon) - (n+1)\epsilon = 1 - n\epsilon > 0. \tag{10.23}$$

Associated with the subset I and $i = 0, \ldots, n$ we define:

$$\begin{aligned} N_i^I &= (1+\epsilon) x_I - [(1+\epsilon)\#I - (n+1)\epsilon] x_i \\ &= (1+\epsilon)\Big(\sum_{j \in I} (x_j - x_i)\Big) + (n+1)\epsilon x_i. \end{aligned} \tag{10.24}$$

From (10.23) and the first formula we get:

$$x_{i_1} \,{}_{(\geq)}\, x_{i_2} \Rightarrow N_{i_1}^I \,{}_{(\leq)}\, N_{i_2}^I. \tag{10.25}$$

That is, strict inequalities are reversed and equalities are preserved.

We can rewrite (10.22) using N_i^I:

$$y_i = \begin{cases} N_i^I / \epsilon[(1+\epsilon)\#I - (n+1)\epsilon] & i \in I \\ -N_i^I / [(1+\epsilon)\#I - (n+1)\epsilon] & i \in \tilde{I} \end{cases} \tag{10.26}$$

The subset I and the values y_i will be determined by the condition $y_i \geq 0$ for all i or equivalently, via (10.26):

$$N_i^I \begin{cases} \geq 0 & i \in I \\ \leq 0 & i \in \tilde{I}. \end{cases} \tag{10.27}$$

Assume that this condition holds. If $x_{i_1} > x_{i_2}$ and $i_1 \in I$ then $N_{i_1}^I \geq 0$ and so, by (10.25), $N_{i_2}^I > 0$ and $i_2 \in I$. That is, (10.27) implies:

$$x_{i_1} > x_{i_2} \text{ and } i_1 \in I \Rightarrow i_2 \in I. \tag{10.28}$$

Now we define a variation of N_i^I which depends only on $i \in \{0, \dots, n\}$

$$\begin{aligned} M_i &= (1+\epsilon)(\sum_{j=0}^{n} \min(x_j, x_i)) - (n+1)x_i \\ &= (1+\epsilon)(\sum_{j=0}^{n}(\min(x_j, x_i) - x_i)) + (n+1)\epsilon x_i. \end{aligned} \tag{10.29}$$

In particular, if $x_{i_0} = \min_{j=0}^{n} x_i$ then $M_{i_0} = (n+1)\epsilon x_{i_0} \geq 0$.

The analogue of (10.25) holds:

$$x_{i_1} \ {}_{(\geq)} x_{i_2} \Rightarrow M_{i_1} \ {}_{(\leq)} M_{i_2}. \tag{10.30}$$

To prove this use the first formula for M_i to see that

$$\begin{aligned} M_{i_1} - M_{i_2} = (1+\epsilon) &\sum_{j=0}^{n}(\min(x_j, x_{i_1}) - \min(x_j, x_{i_2})) \\ &- (n+1)(x_{i_1} - x_{i_2}). \end{aligned} \tag{10.31}$$

If $x_j \leq x_{i_2}$ then the difference of the min's is zero. For the remaining terms, at most n of them, the difference of the min's is bounded by $x_{i_1} - x_{i_2}$. Hence,

$$\begin{aligned} M_{i_1} - M_{i_2} &\leq [(1+\epsilon)n - (n+1)](x_{i_1} - x_{i_2}) \\ &= -[1 - n\epsilon](x_{i_1} - x_{i_2}) \ {}_{(\leq)} 0. \end{aligned} \tag{10.32}$$

Now let I be a nonempty subset of $\{0, \dots, n\}$ with complement $\tilde{I}$. Define:

$$x_{i_I} = \max_{j \in I} x_j$$

$$\tilde{x}_{i_I} = \min_{j \in I} x_j. \tag{10.33}$$

We next prove that if I satisfies (10.28) then

$$N_i^I \geq M_i \text{ with equality if either}$$

$$x_i = x_{i_I} \text{ or } x_i = \tilde{x}_{i_{\tilde{I}}}. \tag{10.34}$$

Assume first $i \in I$. Then by (10.28) $j \in \tilde{I}$ implies $x_j \geq x_i$ and so $\min(x_j, x_i) - x_i = 0$. For $j \in I$ we have $x_j \geq \min(x_j, x_i)$ with equality if $x_i = x_{i_I}$. So using the second pair of formulas in (10.24) and (10.29) we have (10.34) in the $i \in I$ case. Now assume $i \in \tilde{I}$ and so $\tilde{I} \neq \emptyset$ and $\tilde{x}_{i_{\tilde{I}}}$ is defined. By (10.28) $j \in I$ implies $x_j \leq x_i$ and so $\min(x_j, x_i) - x_i = x_j - x_i$. For $j \in \tilde{I}$ we have $\min(x_j, x_i) - x_i \leq 0$ with equality if $x_i = \tilde{x}_{i_{\tilde{I}}}$. Again (10.24) and (10.29) yield (10.34).

We will now show that for a nonempty subset I, the condition (10.27) is equivalent to:

$$I \supset \{i : M_i > 0\}$$

$$\tilde{I} \supset \{i : M_i < 0\}. \tag{10.35}$$

We have seen that (10.27) implies (10.28). We similarly have that (10.35) implies (10.28). The proof uses (10.30) in place of (10.25).

(10.27) $\Rightarrow$ (10.35): Let $i_I \in I$ be such that $x_{i_I} = \max_{j \in I} x_j$. By (10.34) and (10.27) $M_{i_I} = N_{i_I}^I \geq 0$. If $j \in I$ then $x_j \leq x_{i_I}$ implies, by (10.30) $M_j \geq M_{i_I}$ and so $M_j \geq 0$. Let $\tilde{i}_I \in \tilde{I}$ be such that $x_{i\tilde{I}} = \min_{j \in \tilde{I}} x_j$. By (10.34) and (10.27) $M_{\tilde{i}_I} = N_{\tilde{i}_I}^I \leq 0$. If $j \in \tilde{I}$ this $x_j \geq x_{\tilde{i}_I}$ implies by (10.30) $M_j \leq M_{\tilde{i}_I}$ and so $M_j \leq 0$. Hence, (10.27) implies

$$I \subset \{i : M_i \geq 0\}$$

$$\tilde{I} \subset \{i : M_i \leq 0\} \tag{10.36}$$

which is equivalent to (10.35).

(10.35) $\Rightarrow$ (10.27): If $i \in I$ then (10.35) implies (10.36) and so by (10.34) $N_i^I \geq M_i \geq 0$. On the other hand, by (10.36) and (10.34) for $\tilde{i}_I : N_{\tilde{i}_I}^I = M_{\tilde{i}_I} \leq 0$. If $i \in \tilde{I}$ then $x_i \geq x_{\tilde{i}_I}$ implies by (10.25) $N_i^I \leq N_{\tilde{i}_I}^I \leq 0$. This proves (10.27).

Notice that for any x,

$$I_x = \{i : M_i \geq 0\} \tag{10.37}$$

is a nonempty subset of $\{0, 1, \ldots, n\}$ satisfying (10.35). If $\{i : M_i = 0\}$ is empty then I_x is the unique subset satisfying (10.35) and so the coordinates $\{y_i\}$ are uniquely defined. On the other hand, if $M_i = 0$ then with $I = I_x$, $x_i = x_{i_I}$ by (10.30). In particular, the x_i's are all equal for $i \in \{i : M_i = 0\}$. For any I satisfying (10.35), $i \in I$ and $M_i = 0$ implies $x_i = x_{i_I}$, while $i \in \tilde{I}$ and $M_i = 0$ similarly implies $x_i = x_{\tilde{i}_I}$ by (10.30). So by (10.34) we have for all I satisfying (10.35) and all i with $M_i = 0$, $N_i^I = M_i = 0$. So from (10.26) it follows that

$$M_i = 0 \Rightarrow y_i = 0 \tag{10.38}$$

for any I satisfying (10.35).

Assume now that I satisfies (10.35). By (10.36), $I \subset I_x$. Suppose it is a proper subset so that $|\#I_x| = |\#I| + p$ for some $p > 0$. For any $j \in I_x \backslash I$ we have $x_{I_x} = x_I + px_j$ because all the x_j's for $j \in I_x \backslash I \subset \{i : M_i = 0\}$ are equal. For $i \in I_x \backslash I$ we have $y_i = 0$ for both I_x and I by (10.38). To show that y_i for I_x equals y_i for I for the remaining values of i it suffices by (10.22) to show that:

$$\frac{x_I}{(1+\epsilon)\#I - (n+1)\epsilon} = \frac{x_I + px_j}{(1+\epsilon)(\#I + p) - (n+1)\epsilon}. \tag{10.39}$$

After cross multiplication and cancellation this equation is equivalent to $pN_j^I = 0$ which holds since $M_j = 0$ implies $N_j^I = M_j$ by (10.34).

Uniqueness of the coordinates y_i subject to the condition $y_i \geq 0$ for all i, implies the independence and so joinability results. Because for every $x \in \sigma$ there exists a nonempty subset I_x with coordinates nonnegative, it follows that $L(\sigma)$ subdivides σ. If x is in the boundary then with $I = \{i : x_i = 0\}$ and $y_i = x_i$ for all i we obtain a nonnegative solution to (10.19) with I nonempty. By uniqueness this is the $L(\sigma)$ coordinatization of x. Hence, $L(\sigma)$ is a subdivision rel $\partial\sigma$. □

For a simplicial complex K, order the simplices of K so that $\dim \sigma_1 > \dim \sigma_2$ implies σ_1 precedes σ_2. For $z \in K$ the *barycentric stellar subdivision at* z of K is given by (10.11) with $L = b(z).\partial z$. The *barycentric lunar subdivision at* z is given by (10.11) with $L = L(z)$ from (10.14). The *barycentric derived stellar subdivision* of K (or just the barycentric subdivision of K) is obtained by using barycentric stellar subdivisions at each $\sigma \in K$ following the given ordering. Similarly, the *barycentric derived lunar subdivision* (or just the lunar subdivision of K) of K is obtained by using barycentric lunar subdivisions at each $\sigma \in K$ following the given ordering. Notice that at each step the later simplices in the ordering remain unaffected.

A typical simplex of the barycentric subdivision of K is of the form $[b(\sigma_0), \ldots, b(\sigma_k)]$ where $\sigma_0 < \ldots < \sigma_k$ and each is a proper face of its successor. For a similar list $\sigma_0 < \ldots < \sigma_k$ a typical simplex of the linear subdivision is of the form $r(\sigma_0)_{J_0} r(\sigma_1)_{J_1} \ldots r(\sigma_k)_{J_k}$ where $J_0 \subset V(\sigma_0)$, $J_1 \subset V(\sigma_1) \backslash V(\sigma_0), \ldots$ and $J_k \subset V(\sigma_k) \backslash V(\sigma_{k-1})$. In either case, the closed simplex is a subset of the open star $\mathrm{St}(\sigma_0)$ and so each of these subdivisions is proper.

In Figure 10 the top row consists of the stellar and lunar subdivisions

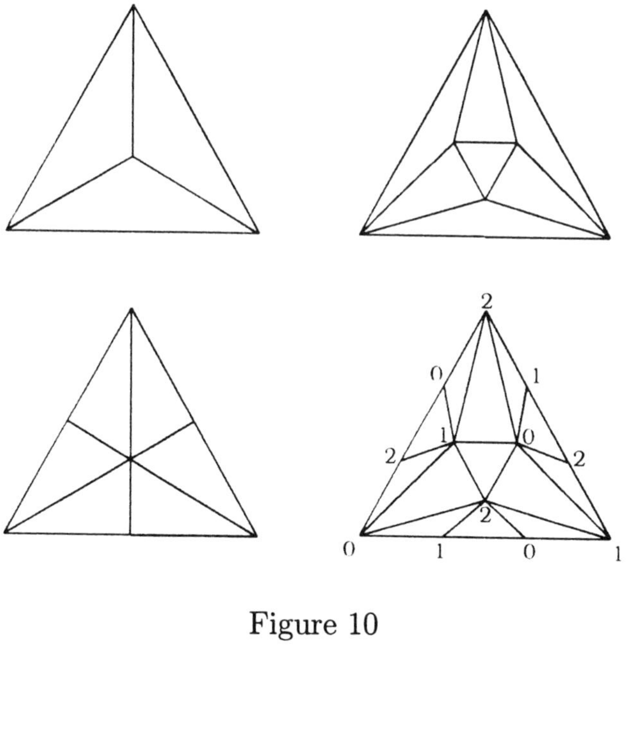

Figure 10

at σ of the two simplex σ. The bottom row consists of the barycentric subdivision and the lunar subdivision of σ. A general *first derived subdivision* of K is a subdivision isomorph of the barycentric derived subdivision. It is obtained by choosing for each $\sigma \in K$ a point $b \in (\sigma)^\circ$ which need not be the barycenter $b(\sigma)$. One can similarly generalize the lunar subdivision.

The lunar subdivision K^* of K is the simplest proper subdivision I know of such that there exists a nondegenerate simplicial map $g : K^* \to K$ whose p.l. map is subordinate to the identity 1_K. Just map each reflection of a vertex back to the original vertex:

$$g(r(\sigma)(v)) = v. \tag{10.40}$$

In the lower right of Figure 10, the lunar subdivision of $\sigma = [0, 1, 2]$ is labeled according to this map.

Recall that if $x = \sum_{i=0}^n x_i v_i$ and $y = \sum_{j=0}^n y_j v_j$ are convex combinations of vectors in a normed linear space, then

$$\begin{aligned} \| x - y \| &= \| \sum_{i,j=0}^{n} x_i y_j (v_i - v_j) \| \\ &\le (1 - \sum_{i=0}^{n} x_i y_i) \max_{i,j=0}^{n} \| v_i - v_j \| . \end{aligned} \tag{10.41}$$

It follows that for K^* a subdivision of K we can define the d_K diameter of a simplex $z^* \in K^*$ in two equivalent ways

$$\begin{aligned} d_K(z^*) &= \sup\{d_K(x_1, x_2) : x_1, x_2 \in z^*\} \\ &= \max\{d_K(v_1, v_2) : v_1, v_2 \in V(z^*)\}. \end{aligned} \tag{10.42}$$

Define the *mesh of* K^* to be

$$d_K(K^*) = \max\{d_K(z^*) : z^* \in K^*\} \tag{10.43}$$

If d is the dimension of K and K^{**} is the barycentric subdivision of K^* then it is easy to check that

$$d_K(K^{**}) \le \frac{d}{d+1} d_K(K^*). \tag{10.44}$$

If K_0^* is a subcomplex of K^* and $z^* \in K^*$ such that $z^* \cap |K_0^*| \ne \emptyset$, i.e. some face of z^* is in K_0^*, then define

$$d_K(z^* : K_0^*) = \sup\{\inf\{d_K(x_1, x_2) :$$

$$x_2 \in z^* \cap |K_0|\} : x_1 \in z^*\}$$

$$= \max\{\inf\{d_K(v_1, x_2) : x_2 \in z^* \cap |K_0^*|\} : v_1 \in V(z^*)\}. \tag{10.45}$$

That is, every point of z^* is at most $d_K(z^* : K_0^*)$ away from some point of a K_0^* face of z^*. If $z^* \cap |K_0^*| = \emptyset$ then define $d_K(z^* : K_0^*) = 1$. Notice that $z^* \in K_0^*$ implies $d_K(z^* : K_0^*) = 0$.

Define the relative mesh to be

$$d_K(K^* : K_0^*) = \max\{\min[d_K(z^*),$$

$$d_K(z^* : K_0^*)] : z^* \in K^*\}. \tag{10.46}$$

2 Proposition. *Let K^* be a subdivision of K and K_0^* be a subcomplex of K^*. Given $\epsilon > 0$ there exists a K^{**} a subdivision of K^* rel K_0^* such that*

$$d_K(K^{**} : K_0^*) < \epsilon. \tag{10.47}$$

Proof. When we say *star on a set of simplices* in a complex we will mean order the set of simplices so that dimension does not increase and then perform successive stellar subdivisions at each simplex following the given order. For example, we initially star at those simplices σ of $K^* \backslash K_0^*$ such that $\partial\sigma \subset K_0^*$. So we can assume that if σ is a K^* simplex which meets K_0^* then $\sigma \cap |K_0^*|$ is a face of σ.

Now let K_1^* be obtained from K^* by starring at each simplex σ of $K^* \backslash K_0^*$ such that $\sigma \cap |K_0^*| \neq \emptyset$. However, choose the cone point b for σ to be not the barycentre of σ but some point of $(\sigma)^\circ$ whose distance from $\sigma \cap |K_0^*|$ is less than ϵ.

Thus, K_1^* is a subdivision of K^* rel K_0^* and for every $z^* \in K_1^*$ such that $z^* \cap |K_0^*| \neq \emptyset$ we have $d_K(z^* : K_0^*) < \epsilon$. Let $\tilde{K}_1 = \{z^* \in K_1^* : z^* \cap |K_0^*| = \emptyset\}$. Define K_2^* by starring K_1^* at the barycenters of the simplices of $\tilde{K}_1$. With $n = 1$ we have the following properties:

(i) K_{n+1}^* is a subdivision of K_n^* rel K_0^*.

(ii) $\tilde{K}_{n+1} = \{z^* \in K^*_{n+1} : z^* \cap |K^*_0| = \emptyset\}$ is the barycentric subdivision of $\tilde{K}_n$.

(iii) If $\sigma \in K^*_n$ is the join $\sigma = \sigma_0\sigma_1$ with $\sigma_0 \in K^*_0$ and $\sigma_1 \in \tilde{K}_n$ then the subdivision induced on σ by K^*_{n+1} is the join of σ_0 with the barycentric subdivision of σ_1.

Proceeding inductively we define K^*_{n+1} by starring K^*_n at the barycenters of $\tilde{K}_n$, and we obtain the above properties.

By (10.44) and (ii) we can choose n so that $d_K(z^*) < \epsilon$ for all $z^* \in \tilde{K}_{n+1}$. By (iii) it follows inductively that for $z^* \in K^*_{n+1}\backslash(\tilde{K}_{n+1} \cup K^*_0)$ $d_K(z^* : K^*_0) < \epsilon$. For suppose that $z^* = \sigma_0\tau \in K^*_{n+1}$ with $\sigma_0 \in K^*_0$ and $\tau \in \tilde{K}_{n+1}$. By (iii) $\tau \subset \sigma_1$ for some $\sigma_1 \in \tilde{K}_n$ such that $\sigma = \sigma_0\sigma_1 \in K^*_n$. Clearly, $d_K(z^* : K^*_0) \leq d_K(\sigma : K^*_0)$. With $K^{**} = K^*_{n+1}$ we then obtain (10.47). □

We conclude with a construction related to that in the above proof. If K is a complex with subcomplex K_0 then a *first derived subdivision rel* K_0 of K is obtained by starring all of the simplices of $K\backslash K_0$ in descending order of dimension. As the name suggests it is a subdivision rel K_0. A typical simplex is of the form $[b(\sigma_1), \ldots, b(\sigma_k)]$ where $\sigma_1 < \ldots < \sigma_k$ is a strictly increasing sequence of simplices of $K\backslash K_0$ or the join of such with $\sigma_0 \in K_0$ where $\sigma_0 < \sigma_1$.

11. Appendix: Hyperbolicity for Relations

Throughout this chapter X is a compact metric space with metric d and F is a closed relation on X, i.e. $F \subset X \times X$. Recall from Chapter 1 that $\overline{V}_\epsilon = \{(x_1, x_2) \in X \times X : d(x_1, x_2) \leq \epsilon\}$. For any subset A, $1_A = \{(x, x) : x \in A\}$, the identity map on A. Using the results of Miller and Akin (1998) we extend the concept of topological hyperbolicity to closed relations and use the sample path spaces to connect the extensions with the homeomorphism concepts described in Akin (1993) Chapter 11 and in Aoki and Hiraide (1994). The latter summarized a broad swath of prior work. For a related approach see Sander (1997).

On the product spaces $X^{\mathbf{Z}}$ and $X^{\mathbf{Z}_+}$ we follow Miller and Akin (1998) defining the metric

$$d(\xi, \eta) = \sup_i \{\min(d(\xi_i, \eta_i), 1/|i|)\} \tag{11.1}$$

with $\min(a, 1/0) \equiv a$ by convention. Thus,

$$d(\xi, \eta) \leq \epsilon \Leftrightarrow$$

$$d(\xi_i, \eta_i) \leq \epsilon \text{ for all } i \text{ such that } |i| < 1/\epsilon. \tag{11.2}$$

We denote by s the shift homeomorphism on $X^{\mathbf{Z}}$ and the shift map on $X^{\mathbf{Z}_+}$ and by π_0 the projection of either of these spaces to X via the 0 coordinate. The canonical projection from $X^{\mathbf{Z}}$ to $X^{\mathbf{Z}_+}$ is denoted π^+. These are all distance nonincreasing and from (11.1) we have

$$\sup_i d(\xi_i, \eta_i) = \sup_i d(s^i(\xi), s^i(\eta)), \tag{11.3}$$

taking the supremum over $i \in \mathbf{Z}$ for $\xi, \eta \in X^{\mathbf{Z}}$ and over $i \in \mathbf{Z}_+$ for $\xi, \eta \in X^{\mathbf{Z}_+}$.

The sample path spaces for F are the closed subsets $X_F \subset X^{\mathbf{Z}}$ and $X_F^+ \subset X^{\mathbf{Z}_+}$ defined by the condition $(\xi_i, \xi_{i+1}) \in F$ for all i, see (1.12). As in Chapter 1 the homeomorphism s_F on X_F and the continuous map s_F^+ on X_F^+ are obtained by restricting the corresponding shift. The restrictions of the projections are denoted $\pi_0 : X_F \to X$, $\pi_0^+ : X_F^+ \to X$ and $\pi_F^+ : X_F \to X_F^+$.

For a closed subset A of X the restriction of F to A is $F_A = F \cap (A \times A)$. The sample path spaces of F_A are

$$A_F = X_F \cap (A^{\mathbf{Z}})$$

(11.4) $$A_F^+ = X_F^+ \cap (A^{\mathbf{Z}_+}).$$

The relation F on X is called *surjective* if $F(X) = X = F^{-1}(X)$. If F is a map then it is surjective as a relation exactly when it is a surjective, i.e. onto, map.

1 Proposition. *Let F be a closed relation on X.*

(11.5) $$\bigcap_{i \in \mathbf{Z}} F^i(X) = \pi_0(X_F)$$

and this subset, denoted $D_\infty(F)$, is called the dynamic domain *of F.*

For a closed subset A of X the following conditions are equivalent and when they hold we call A a surjective subset *of X.*

(1) F_A is a surjective relation on A.

(2) $A \subset F(A) \cap F^{-1}(A)$

(3) $\pi_0(A_F) = A$

(4) There exists an s_F invariant subset K of X_F such that $\pi_0(K) = A$.

The dynamic domain of F is the maximum surjective subset of X, i.e. if A is surjective then $A \subset D_\infty(F)$. In particular, F is surjective iff $D_\infty(F) = X$.

Proof. If $\xi \in X_F$ then $\xi_0 \in F^i(\xi_{-i})$ and so $\xi_0 \in F^i(X)$ for all i. Conversely, suppose x is in the intersection so that there exists for every $n > 0$, x_n and x_{-n} such that $x \in F^n(x_{-n})$ and $x \in F^{-n}(x_n)$. Then we can define $\xi^n \in X^{\mathbf{Z}}$ so that $\xi^n_{-n} = x_{-n}$, $\xi^n_0 = x$, $\xi^n_n = x_n$ and $\xi^n_{i+1} \in F(\xi^n_i)$ for $-n \leq i < n$. If ξ is a limit point of the sequence $\{\xi^n\}$ in $X^{\mathbf{Z}}$ as $n \to \infty$ then $\xi \in X_F$ and $\xi_0 = x$.

This proves (11.5). Clearly, if F is surjective it follows that $\pi_0(X_F) = X$. In particular, applied to F_A we get (1) $\Rightarrow$ (3). The implications (2) $\Leftrightarrow$ (1) and (3) $\Rightarrow$ (4) are obvious. To prove (4) $\Rightarrow$ (1) let $x \in A$ and choose $\xi \in K$ such that $\xi_0 = x$. Since $\xi \in X_F$, $x \in F(\xi_{-1})$ and $x \in F^{-1}(\xi_1)$. But by invariance, $(s_F)^{-1}(\xi)$, $s_F(\xi) \in K$ and so $\xi_{-1} = \pi_0((s_F)^{-1}(\xi))$ and $\xi_1 = \pi_0(s_F(\xi))$ are in $\pi_0(K) \subset A$.

For any subset A of X, $\pi_0(A_F) \subset \pi_0(X_F)$. So if A is surjective we have $A \subset D_\infty(F)$. On the other hand, $D_\infty(F) = \pi_0(X_F)$ satisfies (4) and so is surjective. Since the relation F is surjective exactly when X is a surjective subset we see that F is surjective iff $D_\infty(F) = X$.

Remark. In general, if K is an invariant subset of X_F such that $\pi_0(K) \subset A$ then $K \subset A_F$. That is, A_F is the maximum invariant subset of $(\pi_0)^{-1}(A)$ in X_F. □

By Miller and Akin (1998) Lemma 4.1 the basic sets of F and the chain recurrent set $|\mathcal{C}F|$ are surjective subsets and so they are contained in $D_\infty(F)$. If F is a map then a closed subset A is surjective exactly when it is invariant, i.e. $F(A) = A$.

For any closed subset A of X

$$D_\infty(F_A) = \pi_0(A_F) = \bigcap_{i \in \mathbf{Z}} (F_A)^i(A) = \bigcap_{i \in \mathbf{Z}} (F_A)^i(X) \tag{11.6}$$

is the maximum surjective subset of A.

2 Lemma. *For closed subsets A and B of X the following conditions are equivalent*

(1) $D_\infty(F_A) \subset D_\infty(F_B)$

(2) $D_\infty(F_A) \subset B$

(3) $D_\infty(F_A) = D_\infty(F_{A \cap B})$

(4) $A_F \subset \pi_0^{-1}(B)$

(5) $A_F = (A \cap B)_F$

Proof. (1) $\Rightarrow$ (2): $D_\infty(F_B) \subset B$.

(2) $\Rightarrow$ (4): $\pi_0(A_F) \subset B$ iff $A_F \subset \pi_0^{-1}(B)$.

(4) $\Rightarrow$ (5): B_F is the maximum s_F invariant subset of $\pi_0^{-1}(B)$. Since A_F is invariant, (4) implies $A_F \subset B_F$ and so $A_F = A_F \cap B_F = (A \cap B)_F$.

(5) $\Rightarrow$ (3): Apply π_0.

(3) $\Rightarrow$ (1): Obvious. □

If F_1 is a closed relation on X_1 then a map $h : X_1 \to X$ maps F_1 to F if $h \times h(F_1) \subset F$ or, equivalently,

$$h \circ F_1 \subset F \circ h. \tag{11.7}$$

A *semiconjugacy* h of F_1 to F is a continuous surjective map such that

$$h \circ F_1 = F \circ h, \tag{11.8}$$

(see Akin (1996) Section 5). For functions inclusion implies equality. So a surjective continuous map of a function F_1 to a function F is a semiconjugacy.

If h maps F_1 to F then the induced map on the product spaces, denoted $h_* : X_1^{\mathbf{Z}} \to X^{\mathbf{Z}}$ satisfies

$$h_*(X_{1F_1}) \subset X_F. \tag{11.9}$$

3 Proposition *Let F_1 and F be closed relations on X_1 and X, respectively; let $h : X_1 \to X$ be continuous mapping F_1 to F; and let A_1 and A be closed subsets of X_1 and X, respectively.*

(a) If A_1 is surjective with respect to F_1 and $A = h(A_1)$ then A is surjective with respect to F.

(b) If h is a semiconjugacy from F_1 to F then

$$h_*(X_{1F_1}) = X_F. \tag{11.10}$$

(c) If A is surjective with respect to F, $A_1 = h^{-1}(A)$ and h is a semiconjugacy (or, more generally, (11.10) holds) then

$$h(D_\infty(F_{1A_1})) = A. \tag{11.11}$$

Proof. (a) If $x \in A_1$ and A_1 is surjective then there exist x_{-1} and $x_1 \in A_1$ such that (x_{-1}, x), $(x, x_1) \in F_1$. Hence, $(h(x_{-1}), h(x))$, $(h(x), h(x_1)) \in F_1 \cap (A \times A)$. Thus, A is surjective.

(b) If $(y_0, y_1) \in F$ and $h(x_0) = y_0$ then (11.8) implies there exists x_1 such that $h(x_1) = y_1$ and $(x_0, x_1) \in F_1$. If $\xi \in X_F$ and $n \in \mathbf{Z}_+$, then we can start at ξ_{-n} and proceed inductively forward to define $\eta^n \in X_1^{\mathbf{Z}}$ so that for $i \geq -n$, $h(\eta_i^n) = \xi_i$ and $(\eta_i^n, \eta_{i+1}^n) \in F_1$. If η is a limit point of the sequence $\{\eta^n\}$ then $\eta \in X_{1F_1}$ and $h_*(\eta) = \xi$.

(c) From (11.10) it follows that with $A_1 = h^{-1}(A)$:

$$h_*(A_{1F_1}) = A_F. \tag{11.12}$$

Now apply $\pi_0 : X_F \to X$. Because $\pi_0 \circ h_* = h \circ \pi_0$ and A is surjective:

$$h(D_\infty(F_{1A_1})) = h(\pi_0(A_{1F_1})) =$$

$$\pi_0 h_*(A_{1F_1}) = \pi_0(A_F) = A.$$

$\square$

A closed subset A of X is called *isolated* (rel a closed subset B of X) with respect to F if there exists $\gamma > 0$ such that

$$\xi \in X_F \text{ and } d(\xi_i, A) \leq \gamma \text{ for all } i \in \mathbf{Z}$$

(11.13) $$\Rightarrow \xi_i \in B \text{ for all } i \in \mathbf{Z},$$

where $d(x, A) = \inf\{d(x, y) : y \in A\}$. We call A isolated if A is isolated (rel A).

4 Proposition. *With F a closed relation on X, let A and B be closed subsets of X.*

(a) A is isolated (rel B) with respect to F iff there exists a closed neighborhood U of A such that the following equivalent conditions hold:

(1) $D_\infty(F_U) \subset B$.

(2) $D_\infty(F_U) \subset D_\infty(F_B)$.

(3) $U_F \subset \pi_0^{-1}(B)$.

(4) $U_F \subset B_F$.

(b) The following conditions are equivalent:

(1) A is isolated (rel B) with respect to F.

(2) A is isolated (rel $D_\infty(F_B)$) with respect to F.

(3) $D_\infty(F_A)$ is isolated (rel $D_\infty(F_B)$) with respect to F.

(4) $\pi_0^{-1}(A)$ is isolated (rel $\pi_0^{-1}(B)$) with respect to s_F.

(5) A_F is isolated (rel B_F) with respect to s_F.

(c) Assume F_1 is a closed relation of X_1 and $h : X_1 \to X$ is continuous mapping F_1 to F. Let $A_1 = h^{-1}(A)$ and $B_1 = h^{-1}(B)$. If A is isolated (rel B) with respect to F then A_1 is isolated (rel B_1) with respect to F_1. Conversely, if A_1 is isolated (rel B_1) and h is a semiconjugacy (or more generally if (11.10) holds) then A is isolated (rel B).

Proof. (a) The equivalences are clear from Lemma 2. Condition (11.13) is true iff (1) holds with U the closed γ neighborhood of A, $\overline{V}_\gamma(A)$.

(b) (1) $\Leftrightarrow$ (2): This follows from the equivalence of (1) with (2) in (a).

(2) $\Rightarrow$ (3): If A is isolated (rel B) then any subset of A is isolated (rel B).

(3) $\Rightarrow$ (1): By (a) there exists a closed neighborhood G of $D_\infty(F_A)$ such that $G_F \subset B_F$. Because G is a neighborhood of $D_\infty(F_A)$, (11.6) and compactness imply that for some whole number N

$$\bigcap_{n=-N}^{N} (F_A)^n(A) \subset \text{Int } G \tag{11.14}$$

the interior of G. Now let $\{U_n\}$ be a decreasing sequence of closed neighborhoods of A with intersection A. Then the sequence of closed relation $\{F_{U_n}\}$ decreases to F_A. By Akin (1996) Proposition 1.4 (and induction) we can find a closed neighborhood U of A such that

$$\bigcap_{n=-N}^{N} (F_U)^n(U) \subset \text{Int } G. \tag{11.15}$$

Now let $\xi \in U_F$. By (11.15) we have $\xi_i \in G$ for each $i \in \mathbf{Z}$. That is, $\xi \in G_F$. Hence, $\xi \in B_F$. Thus, we have $U_F \subset B_F$ and so by (a) again A is isolated (rel B).

Before completing the proof of (b), we prove (c).

If A is isolated (rel B) then $U_F \subset B_F$ for some closed neighborhood U of A. Let $U_1 = h^{-1}(U)$, a closed neighborhood of $A_1 = h^{-1}(A)$. If $\xi \in U_{1F_1}$ then by (11.9) $h_*(\xi) \in X_F$ with $h_*(\xi)_i = h(\xi_i) \in U$ for all i, i.e. $h_*(\xi) \in U_F \subset B_F$. So $h(\xi_i) \in B$ for all i, implying $\xi_i \in B_1 = h^{-1}(B)$ for all i.

Now assume A_1 is isolated (rel B_1), i.e. $U_{1F_1} \subset B_{1F_1}$ for some closed neighborhood U_1 of $A_1 = h^{-1}(A)$. By compactness, there exists a closed neighborhood U of A such that $h^{-1}(U) \subset U_1$. If $\xi \in U_F$ and (11.10) holds then there exists $\eta \in X_{1F_1}$ such that $h_*(\eta) = \xi$. $h(\eta_i) = \xi_i \in U$ for all i and so $\eta \in U_{1F_1} \subset B_{1F_1}$. Thus, $\eta_i \in B_1 = h^{-1}(B)$ and $\xi_i = h(\eta_i) \in B$ for all i. That is, $U_F \subset B_F$.

Returning to (b), (1) $\Leftrightarrow$ (4): The continuous function $\pi_0 : X_F \to X$ maps the homeomorphism s_F to the closed relation F but it is usually not a semiconjugacy. It is not even surjective when F is not a surjective relation. On the other hand, if F is a homeomorphism then π_0 is a homeomorphism identifying the shift homeomorphism s_F with the original homeomorphism

F. Applying this remark to the homeomorphism s_F itself we identify its shift space $(X_F)_{s_F}$ with X_F itself. The composed map

$$X_F \cong (X_F)_{s_F} \xrightarrow{\pi_{0*}} X_F \tag{11.16}$$

is the identity map and so (11.10) holds for π_0. In particular, the equivalence of (1) with (4) follows from (c).

(4) $\Leftrightarrow$ (5): A_F is the maximum s_F invariant subset of $\pi_0^{-1}(A)$ and similarly for B_F. Thus, the equivalence of (4) with (5) is just (1) $\Leftrightarrow$ (3) applied to s_F. □

$F \times F$ is the closed relation on $X \times X$ with

$$F \times F(x_1, x_2) = F(x_1) \times F(x_2). \tag{11.17}$$

If A is a closed subset of X then A is surjective with respect to F iff $A \times A$ is surjective with respect to $F \times F$ and iff $1_A = \{(x,x) : x \in A\}$ is surjective with respect to $F \times F$.

A closed subset A of X is called *expansive* for F if 1_A is isolated (rel 1_X) with respect to $F \times F$. That is, there exists $\gamma > 0$ (called the *expansivity constant* for A) such that

$$\begin{gathered} \xi, \eta \in X_F \text{ and } \max(d(\xi_i, A), d(\eta_i, A), d(\xi_i, \eta_i)) \leq \gamma \text{ for all } i \in \mathbf{Z} \\ \Rightarrow \xi_i = \eta_i \quad \text{for all} \quad i \in \mathbf{Z}. \end{gathered} \tag{11.18}$$

F is called an *expansive relation* if X is an expansive subset, i.e. 1_X is isolated with respect to $F \times F$.

5 Proposition. *Let $h : X_1 \to X$ be a semiconjugacy from the closed relation F_1 on X_1 to the closed relation F on X. F is an expansive relation iff $h^{-1} \circ h$ is an isolated subset of $X_1 \times X_1$.*

Proof. The closed equivalence relation $h^{-1} \circ h$ is $(h \times h)^{-1}(1_X)$. F is expansive iff 1_X is isolated with respect to $F \times F$. Since $h \times h$ is a semiconjugacy from $F_1 \times F_1$ to $F \times F$, Proposition 4c implies that 1_X is isolated for $F \times F$ iff $h^{-1} \circ h$ is isolated for $F_1 \times F_1$. □

Let $\gamma \geq 0$. An element ξ of $X^{\mathbf{Z}}$ is called a γ chain for F if it is a chain for the relation $\overline{V}_\gamma \circ F$, i.e.

$$d(\xi_{i+1}, F(\xi_i)) \leq \gamma \quad \text{for all} \quad i \in \mathbf{Z}. \tag{11.19}$$

Elements ξ, η of $X^{\mathbf{Z}}$ are said to γ shadow one another (or ξ γ shadows η) if

$$d(\xi_i, \eta_i) \leq \gamma \text{ for all } i \in \mathbf{Z}. \tag{11.20}$$

If A is a surjective closed subset of X then A satisfies the *Shadowing Property* in X if for every $\epsilon > 0$ there exists a $\delta > 0$ so that any δ chain for F in A is ϵ shadowed by some chain (i.e. 0 chain) in X. That is, if $\xi \in A^{\mathbf{Z}}$ with $d(\xi_{i+1}, F(\xi_i)) \leq \delta$ for all $i \in \mathbf{Z}$, then there exists $\eta \in X_F$ such that $d(\xi_i, \eta_i) \leq \epsilon$ for all $i \in \mathbf{Z}$.

We will need a pair of technical lemmas.

6 Lemma. *Let A be a closed subset of X. For every $\epsilon > 0$ there exists $\delta > 0$ such that every δ chain for F in $\overline{V}_\delta(A)^{\mathbf{Z}}$ is ϵ shadowed by some ϵ chain for F_A.*

Proof. In $A \times A$, the composition $\overline{V}_{\epsilon/2} \circ F_A \circ \overline{V}_{\epsilon/2}$ is a neighborhood of the compact set F_A. As $\delta > 0$ decreases to 0 the compacta $(\overline{V}_\delta \circ F) \cap (\overline{V}_\delta(A) \times \overline{V}_\delta(A))$ decrease to F_A. So there exists $\delta > 0$ such that

$$(\overline{V}_\delta \circ F) \cap (\overline{V}_\delta(A) \times \overline{V}_\delta(A)) \subset \overline{V}_{\epsilon/2} \circ (F_A) \circ \overline{V}_{\epsilon/2}. \tag{11.21}$$

If $\eta \in \overline{V}_\delta(A)^{\mathbf{Z}}$ is a δ chain then (η_i, η_{i+1}) is in $(\overline{V}_\delta \circ F) \cap (\overline{V}_\delta(A) \times \overline{V}_\delta(A))$ and so there exists $\xi_i \in A$ satisfying:

$$d(\eta_i, \xi_i) \leq \epsilon/2$$

$$d(\eta_{i+1}, F_A(\xi_i)) \leq \epsilon/2. \tag{11.22}$$

Hence, $d(\xi_{i+1}, F_A(\xi_i)) \leq d(\xi_{i+1}, \eta_{i+1}) + d(\eta_{i+1}, F_A(\xi_i)) \leq \epsilon$. Thus, $\xi \in A^{\mathbf{Z}}$ is an ϵ chain for F_A and by (11.22) $\epsilon/2$ shadows η. □

7 Corollary. *Let F be a closed relation on X and A be a surjective subset of X. A satisfies the Shadowing Property in X iff for every $\epsilon > 0$ there exists a $\delta > 0$ so that any δ chain for F_A is ϵ shadowed by some 0 chain for F in X. That is, if $\xi \in A^{\mathbf{Z}}$ with $d(\xi_{i+1}, F(\xi_i) \cap A) \leq \delta$ for all $i \in \mathbf{Z}$ then there exists $\eta \in X_F$ such that $d(\xi_i, \eta_i) \leq \epsilon$ for all $i \in \mathbf{Z}$.*

Proof. Assume δ_1 chains for F_A are $\epsilon/2$ shadowed by 0 chains for F. Choose $\delta \leq \delta_1$ so that any δ chain for F in $A^{\mathbf{Z}}$ can be $\epsilon/2$ shadowed by δ_1 chains for F_A. That is, use Lemma 6 with ϵ replaced by $\min(\epsilon/2, \delta_1)$. Then any δ chain for F in $A^{\mathbf{Z}}$ is ϵ shadowed by a 0 chain for F. The converse is obvious. □

8 Lemma. *Let* $0 < \epsilon < 1$.

(a) Assume $\{\xi^i : i \in \mathbf{Z}\}$ *is an* ϵ *chain for* s_F, *i.e.* $\xi^i \in X_F$ *and* $d(\xi^{i+1}, s_F(\xi^i)) \leq \epsilon$ *for all* $i \in \mathbf{Z}$. *If* $\eta_i = \xi_0^i = \pi_0(\xi^i)$ *for* $i \in \mathbf{Z}$, *then* $\eta \in X^{\mathbf{Z}}$ *is an* ϵ *chain for* F *and*

$$d(s^i(\eta), \xi^i) \leq \sqrt{\epsilon} \quad \text{for all } i \in \mathbf{Z}. \tag{11.23}$$

(b) Assume F *is surjective. There exists* δ *with* $0 < \delta \leq \epsilon$ *such that whenever* $\eta \in X^{\mathbf{Z}}$ *is a* δ *chain for* F, *there exists* $\{\xi^i : i \in \mathbf{Z}\}$ *an* ϵ *chain for* s_F *such that*

$$d(s^i(\eta), \xi^i) \leq \epsilon \quad \text{for all } i \in \mathbf{Z}. \tag{11.24}$$

Proof. (a) Because $\xi^i \in X_F$, $\xi_1^i \in F(\xi_0^i)$ and so

$$\begin{aligned} &d(\eta_{i+1}, F(\eta_i)) \leq d(\xi_0^{i+1}, \xi_1^i) = \\ &d(\xi_0^{i+1}, s_F(\xi^i)_0) \leq d(\xi^{i+1}, s_F(\xi^i)) \leq \epsilon. \end{aligned} \tag{11.25}$$

Now let $|j| < 1/\epsilon$:

$$\begin{aligned} &d(s^i(\eta)_j, \xi_j^i) = d(\eta_{i+j}, \xi_j^i) = \\ &d(\xi_0^{i+j}, \xi_j^i) \leq \sum_k d(\xi_{k+1}^{i+j-k-1}, \xi_k^{i+j-k}) \\ &= \sum_k d(s_F(\xi^{i+j-k-1})_k, \xi_k^{i+j-k}) \end{aligned} \tag{11.26}$$

where the summation is over $0 \leq k < j$ if $j > 0$ and over $j \leq k < 0$ if $j < 0$. Because $|k| \leq |j| < 1/\epsilon$, (11.2) implies that each term is bounded by ϵ because $\{\xi^i\}$ is an ϵ chain. Thus, $|j| < 1/\epsilon$ implies

$$d(s^i(\eta)_j, \xi_j^i) \leq |j|\epsilon \quad \text{for all } i \in \mathbf{Z}. \tag{11.27}$$

It follows that $|j| < 1/\sqrt{\epsilon} < 1/\epsilon$ implies $d(s^i(\eta)_j, \xi_j^i) \leq \sqrt{\epsilon}$, yielding (11.23).

(b) Fix $n > 2/\epsilon$. By Lemma 4.2a of Miller and Akin (1998) there exists

$\delta > 0$ such that for every η, a δ chain for the surjective relation F, and $i \in \mathbf{Z}$ there exists $\xi^i \in X_F$ satisfying

$$d(\xi^i_j, \eta_{i+j}) \leq \epsilon/2 \ \text{ for } |j| \leq n+1. \tag{11.28}$$

In particular, for $|j| < 1/\epsilon \leq n$

$$d(s_F(\xi^i)_j, \xi^{i+1}_j) \leq$$

$$d(\xi^i_{j+1}, \eta_{i+j+1}) + d(\xi^{i+1}_j, \eta_{i+j+1}) \leq \epsilon. \tag{11.29}$$

So by (11.2) again $d(s_F(\xi^i), \xi^{i+1}) \leq \epsilon$. That is, $\{\xi^i\}$ is an ϵ chain for s_F. By (11.28)

$$d(\xi^i_j, s^i(\eta)_j) \leq \epsilon \ \text{ for } |j| < 1/\epsilon, \tag{11.30}$$

from which (11.24) follows from (11.2) □

9 Proposition. *Let F be a closed relation on X and A be a surjective subset of X. A satisfies the Shadowing Property for F iff A_F satisfies the Shadowing Property for s_F.*

Proof. Assume A satisfies Shadowing. Given $\epsilon > 0$ with $\epsilon < 1$, let $\epsilon_1 = (\epsilon/2)^2$ and let $\delta > 0$ be such that $\delta < \epsilon_1$ and any δ chain for F in $A^{\mathbf{Z}}$ is ϵ_1 shadowed by some element of X_F.

Let $\{\xi^i : i \in \mathbf{Z}\}$ be a δ chain for s_F in A_F. Define $\eta_i = \xi^i_0$, so that $\eta \in A^{\mathbf{Z}}$. By Lemma 8a, η is a δ chain for F and for all $i \in \mathbf{Z}$,

$$d(s^i(\eta), \xi^i) \leq \sqrt{\delta} \leq \epsilon/2. \tag{11.31}$$

By choice of δ there exists $\zeta \in X_F$ such that $d(\eta_i, \zeta_i) \leq \epsilon_1 \leq \epsilon/2$ for all $i \in \mathbf{Z}$. So by (11.3) we have

$$d(s^i(\zeta), s^i(\eta)) \leq \epsilon/2 \ \text{ for all } i \in \mathbf{Z}. \tag{11.32}$$

By the triangle inequality, $\{s^i_F(\zeta)\}$ is a chain in X_F which ϵ shadows $\{\xi^i\}$.

Now assume A_F satisfies Shadowing for s_F. Given $0 < \epsilon < 1$, let $\epsilon_1 = \epsilon/2$ and choose $\delta_1 > 0$ with $\delta_1 \leq \epsilon_1$ so that any δ_1 chain for s_F in A_F can be ϵ_1 shadowed by some 0 chain for s_F. Because A is a surjective subset the closed relation F_A on A is surjective. Apply Lemma 8b with F on X replaced by F_A on A and ϵ replaced by δ_1. Choose δ so that $0 < \delta \leq \delta_1$

satisfies the conditions of the Lemma. That is, suppose η is a δ chain for F_A. By the choice of δ there exists $\{\xi^i : i \in \mathbf{Z}\}$ a δ_1 chain for s_F such that $\xi^i \in A_F$ and $d(s^i(\eta), \xi^i) \leq \delta_1$ for all $i \in \mathbf{Z}$. By choice of δ_1 there exists $\zeta \in X_F$ with

$$d(s_F^i(\zeta), \xi^i) \leq \epsilon_1 \text{ for all } i \in \mathbf{Z}. \tag{11.33}$$

Thus, ζ is a 0 chain for F and for all $i \in \mathbf{Z}$

$$d(s_F^i(\zeta), s^i(\eta)) \leq \delta_1 + \epsilon_1 \leq \epsilon.$$

By (11.2), ζ ϵ shadows η. By Corollary 7, it follows that A satisfies the Shadowing Property. □

A closed surjective subset A is called a *hyperbolic* subset for F if it is an expansive subset which satisfies the Shadowing Property. This says that there exists $\gamma > 0$ so that for every $\epsilon > 0$ with $\epsilon \leq \gamma$, there exists $\delta > 0$ so that any δ chain for F in A is ϵ shadowed by a unique 0 chain for F in X.

F is called an *Anosov relation* if it is a surjective relation and X is hyperbolic for F.

10 Proposition. *Let F be a closed relation on X and A be a surjective subset of X. The following conditions are equivalent and when they hold we call A an* Anosov subset.

(1) The restriction F_A is an Anosov relation on A and A is an isolated subset of X.

(2) The restriction F_A is an Anosov relation on A and A is an expansive subset for F.

(3) A is an isolated, hyperbolic subset of X.

Proof. (3) $\Rightarrow$ (1) and (2): Let $\gamma > 0$ satisfy (11.13) with $B = A$ and (11.18). Given $\epsilon > 0$, choose $\delta > 0$ so that any δ chain for F in A can be $\min(\epsilon, \gamma)$ shadowed by a 0 chain for F. So if ξ is a δ chain for F_A there exists η in X_F with $d(\xi_i, \eta_i) \leq \min(\epsilon, \gamma)$ for all i. By (11.13) it follows that $\eta_i \in A$ for all i and so $\eta \in A_F$. Thus, η is an F_A chain ϵ shadowing ξ. This implies that A satisfies the Shadowing Property for F_A. A is expansive for F_A with the same constant γ. Thus, F_A is Anosov. A is isolated and expansive for F by assumption.

(1) and (2) $\Rightarrow$ (3): By Corollary 7, A satisfies the Shadowing Property when F_A is Anosov. By assumption A is isolated and expansive for F.

(1) $\Rightarrow$ (2): Let $\gamma > 0$ satisfy (11.13) with $B = A$ and (11.18) for F_A. It follows that (11.18) holds for F. That is, if $\xi, \eta \in X_F$ and $d(\xi_i, A), d(\eta_i, A) \leq \gamma$ for all i, then by (11.13), $\xi_i, \eta_i \in A$ for all i. That is, $\xi, \eta \in A_F$ and so (11.18) for F_A implies $\xi_i = \eta_i$ for all i.

(2) $\Rightarrow$ (1): Let $\gamma > 0$ satisfy (11.18). Choose $0 < \delta_1 \leq \gamma/2$ so that every δ_1 chain for F_A can be $\gamma/2$ shadowed by some F_A chain. By Lemma 5, we can choose $0 < \delta \leq \delta_1$ so that any δ chain for F in $V_\delta(A)^{\mathbf{Z}}$ can be $\gamma/2$ shadowed by a δ_1 chain for F_A. Assume $\xi \in X_F$ with $d(\xi_i, A) \leq \delta$ for all $i \in \mathbf{Z}$. We prove $\xi_i \in A$ for all i which will imply A is isolated. Since ξ is an F chain in $\overline{V}_\delta(A)^{\mathbf{Z}}$ it is $\gamma/2$ shadowed by some δ_1 chain for F_A, η. So η is $\gamma/2$ shadowed by some F_A chain $\tilde{\xi}$. In particular, $\xi, \tilde{\xi} \in X_F$ with $d(\xi_i, \tilde{\xi}_i) \leq \gamma$ for all i and $\tilde{\xi}_i \in A$ for all i. By (11.18) $\xi_i = \tilde{\xi}_i$ and so $\xi_i \in A$ for all i. □

11 Theorem. *Let F be a closed relation on X with s_F the sample path homeomorphism on X_F and s_F^+ the sample path map on X_F^+. Let A be a surjective subset of X. Each of the following properties holds for A with respect to F iff the corresponding property holds for A_F with respect to s_F and iff the corresponding property holds for A_F^+ with respect to s_F^+.*

(1) A is isolated.

(2) A is expansive.

(3) A satisfies the Shadowing Property.

(4) A is hyperbolic.

(5) A is Anosov.

Proof. We first do the comparisons for F and s_F. For (1) we apply Proposition 4b with $A = B$. For (2) we apply Proposition 4b to the relation $F \times F$ and the closed subsets 1_A and 1_X. Observe that $(1_A)_{F\times F} = 1_{A_F}$. For (3) apply Proposition 9. For (4) use (2) and (3). For (5) use (1), (2) and (3), applying Proposition 10.

For the continuous map s_F^+ on X_F^+, the projection $\pi_0^+ : X_F^+ \to X$ induces $(\pi_0^+)_* : X_{s_F^+} \to X_F$, which is a homeomorphism identifying the shift $s_{s_F^+}$ with s_F (see diagram (1.16)). The subset $(A_F^+)_{s_F^+}$ is identified with A_F. Thus, the equivalence between the conditions on A_F^+ with respect to s_F^+ and those on A_F with respect to s_F reduce to the special case of what has already been proved applied to the closed relation s_F^+. □

Now we describe some simple examples.

12 Lemma. *If A is a clopen subset of X, then A is isolated with respect to F. If F is a clopen surjective relation on X then F satisfies the Shadowing Property.*

Proof. If A is open as well as closed then $U = A$ is a neighborhood of A with $D_\infty(F_U) \subset A$. If F is open as well as closed then for some $\epsilon > 0$, $\overline{V}_\epsilon \circ F = F$. So any ϵ chain for F is a 0 chain for F. □

13 Corollary. *(a) If X is any compact metric space then the shift homeomorphism on $X^{\mathbf{Z}}$ and the shift map on $X^{\mathbf{Z}_+}$ satisfy the Shadowing Property.*

(b) If X is a finite set and F is any relation on X then s_F on X_F is an Anosov homeomorphism and s_F^+ on X_F^+ is an Anosov relation.

Proof. In (a) let $F = X \times X$. In both (a) and (b) F is clopen. In (b) we replace X by $D_\infty(F)$ if necessary to assume F is surjective. In either case F satisfies Shadowing and so s_F and s_F^+ do by Theorem 11. In (b) 1_X is a clopen subset and so is isolated. Thus, F is expansive as well. Hence, F and its shifts are hyperbolic. □

Let X^n be the product of n copies of X and define $\pi^n : X^{\mathbf{Z}} \to X^n$ by

$$\pi^n(\xi) = (\xi_0, \ldots, \xi_{n-1}). \tag{11.34}$$

If G is a clopen subset of X^n then $(\pi^n)^{-1}(G)$ a clopen subset of $X^{\mathbf{Z}}$. Its maximum invariant subset, denoted X_G, is therefore isolated with respect to the shift

$$X_G = \{\xi \in Z^{\mathbf{Z}} : \pi^n(s^i(\xi)) \in G \text{ for all } i \in \mathbf{Z}\}. \tag{11.35}$$

We let s_G denote the restriction of s to X_G and call s_G the *subshift of finite type* associated with G. Since s satisfies the Shadowing Property and X_G is isolated it easily follows that s_G satisfies the Shadowing Property.

14 Proposition. *Let X be a finite set and K be a closed subset of $X^{\mathbf{Z}}$*

invariant with respect to the shift homeomorphism s. Let s_K denote the restriction of s to K. The following conditions are equivalent:

(1) *K is isolated with respect to s.*

(2) *K is an Anosov subset, i.e. s_K is an Anosov homeomorphism on K.*

(3) *s_K on K is a subshift of finite type.*

Proof. By Corollary 13, s is an Anosov homeomorphism on $X^{\mathbf{Z}}$. So any closed invariant subset, i.e. closed surjective subset, is hyperbolic with respect to s. So by Proposition 10, (1) $\Leftrightarrow$ (2). We saw above that for a subshift of finite type X_G is isolated and so (3) $\Rightarrow$ (1).

(1) $\Rightarrow$ (3): If K is isolated then it has a closed neighborhood U such that $K = \cap_{i=-\infty}^{\infty} s^{-i}(U)$. Because $X^{\mathbf{Z}}$ is zero-dimensional the clopen sets form a basis for the topology and so we can assume U is clopen. Any clopen subset is the finite union of basic open sets. That is, there exists a subset G of the product of $2n+1$ copies of X for some n, so that

$$U = \{\xi : (\xi_{-n}, \dots, \xi_n) \in G\}. \tag{11.36}$$

Since K is invariant we can shift and see that $K = \cap_{i=-\infty}^{\infty} s^{-i}(\tilde{U})$ where $\tilde{U} = (\pi^{2n+1})^{-1}(G)$. That is, s_K is the subshift associated with G.

Remark. We also call the invariant set K itself a subshift of finite type. □

15 Proposition. *Let f_1 be a homeomorphism on X_1. For X a finite set let K be a closed subset of $X^{\mathbf{Z}}$ invariant with respect to s, such that the restriction s_K is a subshift of finite type. Assume $h : K \to X_1$ is surjective and continuous mapping s_K to f_1. The following conditions are equivalent:*

(1) *f_1 is an expansive map on X_1.*

(2) *$h^{-1} \circ h$ is an isolated invariant subset of $K \times K$.*

(3) *With respect to the shift $s \times s$ on $X^{\mathbf{Z}} \times X^{\mathbf{Z}}$, i.e. s on $(X \times X)^{\mathbf{Z}}$, $h^{-1} \circ h$ is a subshift of finite type.*

Proof. Since f_1 and s_K are maps and h is surjective, h is a semiconjugacy. Since f_1 is a homeomorphism then subset $(h \times h)^{-1}(1_{X_1}) = h^{-1} \circ h$ is an invariant subset of $K \times K$.

(2) $\Leftrightarrow$ (3): Since $K \times K$ is isolated in $X^{\mathbf{Z}} \times X^{\mathbf{Z}}$, it follows easily that $h^{-1} \circ h$ is isolated in $K \times K$ iff it is isolated in $X^{\mathbf{Z}} \times X^{\mathbf{Z}}$. The equivalence then follows from Proposition 14.

(1) $\Leftrightarrow$ (2): Apply Proposition 5. $\square$

References

E. Akin, *Manifold phenomena in the theory of polyhedra*, Trans. Amer. Math. Soc., (1969) 143: 413-473.

____, *The general topology of dynamical systems*, Amer. Math. Soc., Providence, 1993.

____, *Dynamical systems: the topological foundations*, in Six Lectures on Dynamical Systems, (B. Aulback and F. Colonius, eds.), pp.1-43, World Scientific, Singapore, 1996.

____, *Recurrence in topological dynamics: Furstenberg families and Ellis actions*, Plenum Press, New York, 1997.

E. Akin, M. Hurley and J. Kennedy, *The generic homeomorphism is complicated but not chaotic*, (1996) to appear.

N. Aoki and K. Hiraide, *Topological theory of dynamical systems: recent advances*, Elsevier Science B.V., Amsterdam, 1994.

M.A. Armstrong, *Transversality for polyhedra*, Ann. of Math., (1967) 86: 172-191.

J. Auslander, *Minimal flows and their extensions*, Elsevier Science B.V., Amsterdam, 1988.

M. Barnsley, *Fractals everywhere*, Academic Press, San Diego, 1988.

P. Billingsley, *Ergodic theory and information*, John Wiley, New York, 1965.

M.M. Cohen, *A general theory of relative regular neighborhoods*, Trans Amer. Math. Soc. (1969) 136: 189-230.

C. C. Conley, *Isolated invariant sets and the Morse index*, CBMS Conf. Series in Math., vol. 38, Amer. Math. Soc., Providence, 1978.

F. R. Gantmacher, *The theory of matrices*, Chelsea, New York, 1959.

L.W. Goodwyn, *Comparing topological entropy with measure theoretical entropy.* Amer. J. of Math. (1972) 94: 366-387.

J. F. P. Hudson, *Piecewise linear topology*, Benjamin, New York, 1969.

C. S. Hsu, *Cell-to-cell mapping: a method of global analysis for nonlinear systems*, Springer-Verlag, Berlin, 1987.

A. Katok and B. Hasselblatt, *Introduction to the modern theory of dynamical systems*, Cambridge Univ. Press, Cambridge, 1995.

B.B. Mandelbrot and S. Jaffard, *Peano-Pólya motions, when time is intrinsic or binomial (uniform or multifractal)*, The Math. Intelligencer (1997) 19: 21-26.

W. Miller and E. Akin, *Invariant measures for set-valued dynamical systems*, Trans. Amer. Math. Soc., (1999), 351: 1203-1225.

J.W. Milnor, *Fubini foiled: Katok's paradoxical example in measure theory*, The Math. Intelligencer (1997) 19: 30-32.

M. Misiurewicz, *A short proof of the variational principle for a* $\mathbf{Z}_+^N$ *action on a compact space*, Astérisque (1976) 40: 147-181.

G. Osipenko, *Symbolic analysis of the chain recurrent trajectories of dynamical systems*, (to appear).

J. Oxtoby, *Measure and category* (2nd ed.), Springer-Verlag, Berlin, 1980.

R. S. Palais, *Foundations of global nonlinear analysis*, Benjamin, New York, 1968.

C. Robinson, *Dynamical systems: stability, symbolic dynamics and chaos*, CRC, Boca Raton, 1995.

C. P. Rourke and B. J. Sanderson, *Introduction to piecewise-linear topology*, Springer-Verlag, Berlin, 1972.

D.J. Rudolph, *Fundamentals of measurable dynamics: ergodic theory on Lebesgue spaces*, Oxford Univ. Press, Oxford, 1990.

E. Sander, *Hyperbolic sets for noninvertible maps and relations*, (to appear).

J. R. Stallings, *Lectures on polyhedral topology*, Lect. Notes on Mathematics, Tata Institute of Fundamental Research, Bombay, 1968.

Index

Editorial Information

To be published in the *Memoirs*, a paper must be correct, new, nontrivial, and significant. Further, it must be well written and of interest to a substantial number of mathematicians. Piecemeal results, such as an inconclusive step toward an unproved major theorem or a minor variation on a known result, are in general not acceptable for publication. *Transactions* Editors shall solicit and encourage publication of worthy papers. Papers appearing in *Memoirs* are generally longer than those appearing in *Transactions* with which it shares an editorial committee.

As of March 31, 1999, the backlog for this journal was approximately 4 volumes. This estimate is the result of dividing the number of manuscripts for this journal in the Providence office that have not yet gone to the printer on the above date by the average number of monographs per volume over the previous twelve months, reduced by the number of issues published in four months (the time necessary for preparing an issue for the printer). (There are 6 volumes per year, each containing at least 4 numbers.)

A Copyright Transfer Agreement is required before a paper will be published in this journal. By submitting a paper to this journal, authors certify that the manuscript has not been submitted to nor is it under consideration for publication by another journal, conference proceedings, or similar publication.

Information for Authors and Editors

Memoirs are printed by photo-offset from camera copy fully prepared by the author. This means that the finished book will look exactly like the copy submitted.

The paper must contain a *descriptive title* and an *abstract* that summarizes the article in language suitable for workers in the general field (algebra, analysis, etc.). The *descriptive title* should be short, but informative; useless or vague phrases such as "some remarks about" or "concerning" should be avoided. The *abstract* should be at least one complete sentence, and at most 300 words. Included with the footnotes to the paper, there should be the 1991 *Mathematics Subject Classification* representing the primary and secondary subjects of the article. This may be followed by a list of *key words and phrases* describing the subject matter of the article and taken from it. A list of the numbers may be found in the annual index of *Mathematical Reviews*, published with the December issue starting in 1990, as well as from the electronic service e-MATH [**telnet e-MATH.ams.org** (or **telnet 130.44.1.100**). Login and password are **e-math**]. For journal abbreviations used in bibliographies, see the list of serials in the latest *Mathematical Reviews* annual index. When the manuscript is submitted, authors should supply the editor with electronic addresses if available. These will be printed after the postal address at the end of each article.

Electronically prepared papers. The AMS encourages submission of electronically prepared papers in $\mathcal{A}_{\mathcal{M}}\mathcal{S}$-TeX or $\mathcal{A}_{\mathcal{M}}\mathcal{S}$-LaTeX. The Society has prepared author packages for each AMS publication. Author packages include instructions for preparing electronic papers, the *AMS Author Handbook*, samples, and a style file that generates the particular design specifications of that publication series for both $\mathcal{A}_{\mathcal{M}}\mathcal{S}$-TeX and $\mathcal{A}_{\mathcal{M}}\mathcal{S}$-LaTeX.

Authors with FTP access may retrieve an author package from the Society's Internet node `e-MATH.ams.org` (130.44.1.100). For those without FTP access, the author package can be obtained free of charge by sending e-mail to `pub@ams.org` (Internet) or from the Publication Division, American Mathematical Society, P.O. Box 6248, Providence, RI 02940-6248. When requesting an author package, please specify $\mathcal{A}\mathcal{M}\mathcal{S}$-TeX or $\mathcal{A}\mathcal{M}\mathcal{S}$-LaTeX, Macintosh or IBM (3.5) format, and the publication in which your paper will appear. Please be sure to include your complete mailing address.

Submission of electronic files. At the time of submission, the source file(s) should be sent to the Providence office (this includes any TeX source file, any graphics files, and the DVI or PostScript file).

Before sending the source file, be sure you have proofread your paper carefully. The files you send must be the EXACT files used to generate the proof copy that was accepted for publication. For all publications, authors are required to send a printed copy of their paper, which exactly matches the copy approved for publication, along with any graphics that will appear in the paper.

TeX files may be submitted by email, FTP, or on diskette. The DVI file(s) and PostScript files should be submitted only by FTP or on diskette unless they are encoded properly to submit through e-mail. (DVI files are binary and PostScript files tend to be very large.)

Files sent by electronic mail should be addressed to the Internet address `pub-submit@ams.org`. The subject line of the message should include the publication code to identify it as a Memoir. TeX source files, DVI files, and PostScript files can be transferred over the Internet by FTP to the Internet node `e-math.ams.org` (130.44.1.100).

Electronic graphics. Figures may be submitted to the AMS in an electronic format. The AMS recommends that graphics created electronically be saved in Encapsulated PostScript (EPS) format. This includes graphics originated via a graphics application as well as scanned photographs or other computer-generated images.

If the graphics package used does not support EPS output, the graphics file should be saved in one of the standard graphics formats—such as TIFF, PICT, GIF, etc.—rather than in an application-dependent format. Graphics files submitted in an application-dependent format are not likely to be used. No matter what method was used to produce the graphic, it is necessary to provide a paper copy to the AMS.

Authors using graphics packages for the creation of electronic art should also avoid the use of any lines thinner than 0.5 points in width. Many graphics packages allow the user to specify a "hairline" for a very thin line. Hairlines often look acceptable when proofed on a typical laser printer. However, when produced on a high-resolution laser imagesetter, hairlines become nearly invisible and will be lost entirely in the final printing process.

Screens should be set to values between 15% and 85%. Screens which fall outside of this range are too light or too dark to print correctly.

Any inquiries concerning a paper that has been accepted for publication should be sent directly to the Editorial Department, American Mathematical Society, P. O. Box 6248, Providence, RI 02940-6248.

Editors

This journal is designed particularly for long research papers (and groups of cognate papers) in pure and applied mathematics. Papers intended for publication in the *Memoirs* should be addressed to one of the following editors:

Ordinary differential equations, partial differential equations, and applied mathematics to JOHN MALLET-PARET, Division of Applied Mathematics, Brown University, Providence, RI 02912-9000; electronic mail: `jmp@cfm.brown.edu`.

Harmonic analysis, representation theory, and Lie theory to ROBERT J. STANTON, Department of Mathematics, The Ohio State University, 231 West 18th Avenue, Columbus, OH 43210-1174; electronic mail: `stanton@math.ohio-state.edu`.

Ergodic theory and dynamical systems to ROBERT F. WILLIAMS, Department of Mathematics, University of Texas at Austin, Austin, TX 78712-1082; e-mail: `bob@math.utexas.edu`

Real and harmonic analysis and geometric partial differential equations to WILLIAM BECKNER, Department of Mathematics, University of Texas at Austin, Austin, TX 78712-1082; e-mail: `beckner@math.utexas.edu`.

Algebra to CHARLES CURTIS, Department of Mathematics, University of Oregon, Eugene, OR 97403-1222 e-mail: `cwc@darkwing.uoregon.edu`

Algebraic topology and cohomology of groups to STEWART PRIDDY, Department of Mathematics, Northwestern University, 2033 Sheridan Road, Evanston, IL 60208-2730; e-mail: `s_priddy@math.nwu.edu`.

Differential geometry and global analysis to CHUU-LIAN TERNG, Department of Mathematics, Northeastern University, Huntington Avenue, Boston, MA 02115-5096; e-mail: `terng@neu.edu`.

Probability and statistics to RODRIGO BAÑUELOS, Department of Mathematics, Purdue University, West Lafayette, IN 47907-1968; e-mail: `banuelos@math.purdue.edu`.

Combinatorics and Lie theory to PHILIP J. HANLON, Department of Mathematics, University of Michigan, Ann Arbor, MI 48109-1003; e-mail: `hanlon@math.lsa.umich.edu`.

Logic to THEODORE SLAMAN, Department of Mathematics, University of California at Berkeley, Berkeley, CA 94720-3840; e-mail: `slaman@math.berkeley.edu`.

Number theory and arithmetic algebraic geometry to ALICE SILVERBERG, MSRI, 1000 Centennial Dr., Berkeley, CA 94720; e-mail: `silver@math.ohio-state.edu`.

Complex analysis and complex geometry to DANIEL M. BURNS, Department of Mathematics, University of Michigan, Ann Arbor, MI 48109-1003; e-mail: `dburns@math.lsa.umich.edu`.

Algebraic geometry and commutative algebra to LAWRENCE EIN, Department of Mathematics, University of Illinois, 851 S. Morgan (M/C 249), Chicago, IL 60607-7045; e-mail: `ein@uic.edu`.

Geometric topology, knot theory, hyperbolic geometry, and general topoogy to JOHN LUECKE, Department of Mathematics, University of Texas at Austin, Austin, TX 78712-1082; e-mail: `luecke@math.utexas.edu`.

Partial differential equations and applied mathematics to BARBARA LEE KEYFITZ, Department of Mathematics, University of Houston, 4800 Calhoun, Houston, TX 77204-3476; e-mail: `keyfitz@uh.edu`

Operator algebras and functional analysis to BRUCE E. BLACKADAR, Department of Mathematics, University of Nevada, Reno, NV 89557; e-mail: `bruceb@math.unr.edu`

All other communications to the editors should be addressed to the Managing Editor, PETER SHALEN, Department of Mathematics, University of Illinois, 851 S. Morgan (M/C 249), Chicago, IL 60607-7045; e-mail: `shalen@math.uic.edu`.

Selected Titles in This Series

(Continued from the front of this publication)

638 **Gunnar Fløystad,** Higher initial ideals of homogeneous ideals, 1998

637 **Thomáš Gedeon,** Cyclic feedback systems, 1998

636 **Ching-Chau Yu,** Nonlinear eigenvalues and analytic-hypoellipticity, 1998

635 **Magdy Assem,** On stability and endoscopic transfer of unipotent orbital integrals on p-adic symplectic groups, 1998

634 **Darrin D. Frey,** Conjugacy of Alt_5 and $\mathrm{SL}(2,5)$ subgroups of $\mathrm{E}_8(\mathbb{C})$, 1998

633 **Dikran Dikranjan and Dmitri Shakhmatov,** Algebraic structure of pseudocompact groups, 1998

632 **Shouchuan Hu and Nikolaos S. Papageorgiou,** Time-dependent subdifferential evolution inclusions and optimal control, 1998

631 **Ronnie Lee, Steven H. Weintraub, and J. William Hoffman,** The Siegel modular variety of degree two and level four/Cohomology of the Siegel modular group of degree two and level four, 1998

630 **Florin Rădulescu,** The Γ-equivariant form of the Berezin quantization of the upper half plane, 1998

629 **Richard B. Sowers,** Short-time geometry of random heat kernels, 1998

628 **Christopher K. McCord, Kenneth R. Meyer, and Quidong Wang,** The integral manifolds of the three body problem, 1998

627 **Roland Speicher,** Combinatorial theory of the free product with amalgamation and operator-valued free probability theory, 1998

626 **Mikhail Borovoi,** Abelian Galois cohomology of reductive groups, 1998

625 **George Xian-Zhi Yuan,** The study of minimax inequalities and applications to economies and variational inequalities, 1998

624 **P. Deift and K. T-R McLaughlin,** A continuum limit of the Toda lattice, 1998

623 **S. A. Adeleke and Peter M. Neumann,** Relations related to betweenness: Their structure and automorphisms, 1998

622 **Luigi Fontana, Steven G. Krantz, and Marco M. Peloso,** Hodge theory in the Sobolev topology for the de Rham complex, 1998

621 **Gregory L. Cherlin,** The classification of countable homogeneous directed graphs and countable homogeneous n-tournaments, 1998

620 **Victor Guba and Mark Sapir,** Diagram groups, 1997

619 **Kazuyoshi Kiyohara,** Two classes of Riemannian manifolds whose geodesic flows are integrable, 1997

618 **Karl H. Hofmann and Wolfgang A. F. Ruppert,** Lie groups and subsemigroups with surjective exponential function, 1997

617 **Robin Hartshorne,** Families of curves in $\mathbb{P}^3$ and Zeuthen's problem, 1997

616 **Serguei G. Bobkov and Christian Houdré,** Some connections between isoperimetric and Sobolev-type inequalities, 1997

615 **Michael A. Dritschel and Hugo J. Woerdeman,** Model theory and linear extreme points in the numerical radius unit ball, 1997

614 **Richard Warren,** The structure of k-CS-transitive cycle-free partial orders, 1997

613 **D. L. Flannery,** The finite irreducible linear 2-groups of degree 4, 1997

612 **Joan Porti,** Torsion de Reidemeister pour les variétés hyperboliques, 1997

611 **D. Ginzburg, I. Piatetski-Shapiro, and S. Rallis,** L functions for the orthogonal group, 1997

610 **Mark Hovey, John H. Palmieri, and Neil P. Strickland,** Axiomatic stable homotopy theory, 1997

609 **Liviu I. Nicolaescu,** Generalized symplectic geometries and the index of families of elliptic problems, 1997

(See the AMS catalog for earlier titles)